Quality Operations Procedures for Pharmaceutical, API, and Biotechnology

Syed Imtiaz Haider
Erfan Syed Asif

CRC Press
Taylor & Francis Group
Boca Raton London New York

CRC Press is an imprint of the
Taylor & Francis Group, an **informa** business

CRC Press
Taylor & Francis Group
6000 Broken Sound Parkway NW, Suite 300
Boca Raton, FL 33487-2742

Printed in the United States of America on acid-free paper
Version Date: 20120413

International Standard Book Number: 978-1-4398-8690-8 (Hardback)

Library of Congress Cataloging-in-Publication Data

Haider, Syed Imtiaz.
Quality operations procedures for pharmaceutical, API, and biotechnology / Syed Imtiaz Haider, Erfan Syed Asif.
p. ; cm.
Includes bibliographical references and index.
ISBN 978-1-4398-8690-8 (hardcover : alk. paper)
I. Asif, Erfan Syed. II. Title.
[DNLM: 1. Pharmaceutical Preparations--standards. 2. Biotechnology--standards. 3. Drug Industry--standards. 4. Quality Control. 5. Technology, Pharmaceutical--standards. QV 771]

615.1--dc23 2012013735

Visit the Taylor & Francis Web site at
http://www.taylorandfrancis.com

and the CRC Press Web site at
http://www.crcpress.com

We dedicate this work to (the late) Prof. Dr. Salimuzzaman Siddiqui, FRS, and Prof. Dr. Atta-ur-Rahman, FRS, PhD, ScD, the two leading Pakistani scientists in natural products chemistry who revolutionized the research on pharmacology and pharmacognosy of various plants to extract novel chemical substances of medicinal importance.

We also acknowledge the affection and support of our families (Haiders and Asifs), especially our wives, who have always played a vital role in our success.

Contents

Preface

Pharmaceutical, medical, and biotech industries are regulated worldwide to be in compliance with current good manufacturing practice (cGMP) and good laboratory practice (GLP) principles. Each company is required to create quality standard operating procedures (SOPs) to build the regulatory requirements into actions and describe the personnel flow, internal flow, flow of information, and processing steps. The template SOPs available enable end users to understand principles and elements of GMP and provide documentation language ranging from generic to specific, depending on the detail level of the requirements.

This book and CD-ROM take into account all major international regulations, such as FDA, EU GMP, cGMP, GLP, PDA technical monographs, PDA technical reports, PMA concepts, journals of PDA, GCP, and industry standard ISO 9000, to be in compliance with documentation guidelines. No other book in print deals exclusively with the key elements of quality control (QC) and quality assurance (QA) procedures for pharmaceutical operations and provides hands-on templates to be tailored to achieve global regulatory compliance.

Quality operations procedures (QOPs) are written to provide explicit instruction on how to achieve the standards for those responsible for writing and executing QC and QA SOPs, procedures for drug, drug–device combinations, diagnostics, pharmaceutical, biotechnology, and bulk pharmaceutical chemicals. Included are ready-to-use templates so that one can immediately save time and expense without missing any critical elements.

The book provides instant answers to quality professionals, QA auditors, and protocol writers about what to include in critical QA and QC SOPs and how to enhance productivity.

The QOPs on the CD-ROM are valuable tools for companies in the process of developing QA and QC systems and organizing SOPs to achieve FDA, GMP, and GLP compliance.

The CD-ROM contains 19 QC and 33 QA SOPs; these documents are readily available as downloadable PDFs so that customers can use them for their own businesses. The contents of the procedures are written in simple and precise language to be in compliance with FDA, GMP, and GLP requirements. The book ensures minimization of the number of documents to avoid the head of QA's nightmare at the time of an FDA audit. The SOPs exclusively refer to the documents specially required for compliance; however, specific formats are not included to ensure that the electronic templates can be easily used by pharmaceutical, bulk pharmaceutical, medical device, and biotechnology industries.

The purpose of the book and CD-ROM is to meet the need for a ready-to-use text on the quality systems of aseptic pharmaceutical nonaseptic production, and to provide general information and guidelines.

The formats and style provided are generic and can be further amended. The contents of the SOPs are intended to build quality and regulatory requirements in QA and QC procedures. However, having a set of validation SOPs does not preclude adverse inspection findings because contents that satisfy one inspector may not satisfy another.

I strongly believe that technical management and staff should read the procedures to ensure that particular needs are addressed with reference to QA and QC within the organization and individual countries' regulatory requirements. It is, however, guaranteed to provide management with a tool to develop a set of quality SOPs to support the road map established for the on-time successful start-up of the facility operation in compliance with the GMP requirements.

There are several key benefits involved in compliance to QA and QC procedures, including but not limited to

- Successful facility operational start-up
- Minimized noncompliance
- Reduced reworks
- Reduced rejected lots
- Avoidance of recalled lots
- Help in new drug approval
- Satisfactory inspections
- Corporate image
- Financial gain
- Secure third-party contracts
- Corporate legal protection
- Fewer complaints
- Reduced testing
- Improved employee awareness

We believe that by following the broad-based example of these SOPs, new as well as experienced companies can benefit by enhancing their existing documentation to meet FDA and other regulatory requirements.

Acknowledgments

We take pride in acknowledging our families for their patience, love, and cooperation without which the accomplishment of this book would have been next to impossible.

Authors

Syed Imtiaz Haider earned his PhD in chemistry and is a quality assurance and environmental specialist with over 20 years of experience in aseptic and nonaseptic pharmaceutical processes, equipment validation, and in-process control and auditing. Dr. Haider is currently involved in several major biotechnology-based tasks, including cell-line qualification, process validation, biological analysis, method validation, biosimilars comparative studies, organizing preclinical studies, and preparation of the Central Technical Dossier (CTD) formatted for regulatory submission. He is the author and coauthor of more than 20 research publications in international refereed journals dealing with products of pharmaceutical interest, their isolation, and structure development. As a professional technical writer, he has authored more than 2000 standard operating procedures (SOPs) based on FDA regulations, ISO 9001:2000, and ISO 14001:2004 standards. He is a certified QMS (Quality Management System) auditor of the IRCA (International Register of Certificated Auditors) and a registered associate environmental auditor for the EARA (Environmental Auditors Registration Association). He has written more than 10 quality system manuals for multidisciplinary industries and provided consultancy to the Drug Control Laboratory of the Ministry of Health in the United Arab Emirates in developing a quality management system based on ISO 9003 and later transition to ISO 9001:2000.

Dr. Haider works as a quality affairs director at Julphar, Gulf Pharmaceutical Industries, and is involved in the preparation of several abbreviated new drug application (ANDA) files, which after successful FDA, EU, and GMP inspections, will lead to the export of finished pharmaceutical products to the United States and European markets. He has also written *ISO 9001:2000: Document Development Compliance Manual: A Complete Guide and CD-ROM; Pharmaceutical Validation Master Plan: The Ultimate Guide to FDA, GMP, GLP Compliance*; and *Validation Standard Operating Procedures and Biotechnology: A Comprehensive Training Guide for the Biotechnology Industry*. Dr. Haider holds the intellectual copyright certificate of registration for an electronic documentation package on ISO 9000 and ISO 14001 from the Canadian Intellectual Property Office. He is also a contributing author of chapters on ISO 9001:2000 and ISO 14001 in international publications.

Dr. Haider has organized cGMP and biotechnology updates and conferences in the region, resourcing competitive speakers from Europe, Canada, and the United States. More recently, he has contributed *A Comprehensive Training Guide for the Biotechnology Industry* and *Cleaning Validation Master Plan with CD ROM*, published by Taylor & Francis Group, CRC Press.

Erfan Syed Asif earned his PhD in organic chemistry and has expertise in various areas of quality operations with over 17 years of experience in pharmaceutical industries in Pakistan, the United States, Canada, and the UAE. He has worked in US FDA– and Health Canada–approved facilities in managerial positions under the quality operations umbrella. He has extensive experience overseeing qualification projects for manufacturing equipments, utilities, systems, sterilization techniques, aseptic processes simulation testing, and the sterile and nonsterile products manufacturing process. Dr. Asif currently holds the position of quality assurance manager in Gulf Pharmaceutical Industries, where his responsibilities include the administrative routine related to quality in the manufacturing areas, investigating and responding to market complaints, advisory to production, writing of SOPs, introduction of new products, and conducting validation of manufacturing processes, cleaning procedures, and equipment and facilities qualification. He is also responsible for overseeing the vendor approval process and for conducting external audits of the API manufacturers for Gulf Pharmaceutical industries (Julphar).

As a validation consultant, he provided training and guidance on projects at Pharmacia & Upjohn, Michigan (USA), GlaxoSmithKline (Canada), and Air Liquide Canada (medical gas manufacturers) based on Health Canada and US FDA regulations. He was also engaged in providing trainings on validation to various groups of validation consultants working on different projects in Canada and the United States.

Dr. Asif is author of many research publications in internationally published chemistry journals and conference proceedings. He has coauthored with Syed Imitiaz Haider *Cleaning Validation Master Plan* and *Quality Control Training Manual*, published by Taylor & Francis Group in 2010 and 2011, respectively. He is also a contributing author of three chapters in the book *Biotechnology Training Manual* by Syed Imitiaz Haider and Anika Ashok, published by Taylor & Francis Group in 2009.

Dr. Asif is also involved, along with the Quality Affairs Director, Julphar, in organizing cGMP conferences in the region, managing to resource competitive speakers from Europe, Canada, and the United States. He is a regular appointee of the Board of Advanced Research and Studies of Karachi University, Pakistan, as an external examiner for thesis evaluation and for conducting viva voce for MPhil and PhD degrees in pharmaceutical chemistry.

QCS-001

YOUR COMPANY STANDARD OPERATING PROCEDURE

SOP No. QCS-001.00 Effective date: mm/dd/yyyy
Approved by:

TITLE: **Handling of Sampled Material/Product**

AUTHOR: ________________________________

Name/Title/Department

____

Signature/Date

CHECKED BY: ________________________________

Name/Title/Department

____

Signature/Date

APPROVED BY: ________________________________

Name/Title/Department

____

Signature/Date

REVISIONS:

No.	Section	Pages	Initials/Date

SOP No. QCS-001.00 Effective date: mm/dd/yyyy
Approved by:

Subject: Handling of Sampled Material/Product

1.1 PURPOSE

To provide written instructions to clarify the steps followed after sampling till the sample reaches the analyst.

1.2 RESPONSIBILITY

It is the responsibility of the quality assurance (QA) inspector and quality control (QC) administration to follow the procedure. QC/QA directors will be responsible for the standard operating procedures (SOP) compliance.

1.3 PROCEDURE

Just after sampling of raw/packaging materials or products (intermediate, bulk, or finished), sampled items are sent to the QC laboratory for further analysis.

- The QA inspector will then enter information regarding the products' (intermediate, bulk, or finished) raw and packaging material in the relevant log book.
- The QA inspector will further forward the samples to the QC laboratory together with relevant documents such as sampling and receiving reports for raw and packaging material, and batch records for intermediate, bulk, or finished products also with sampling report.
- The QC sample receiving personnel will verify samples and documents and will enter the sample information in a relevant analysis journal.
- The receiving personnel will then assign a QC number for each sample and distribute samples to relevant sections of the laboratory supervisor or officer (physical, chemical, and/or microbiological) as per the material/product analytical specification.

For not more than one hour should the samples stay in the receiving room before they reach the corresponding QC supervisor or officer.

1.4 REASON FOR REVISION

First time issued for your company, affiliates, and contract manufacturers.

QCS-002

YOUR COMPANY STANDARD OPERATING PROCEDURE

SOP No. QCS-002.00 Effective date: mm/dd/yyyy
Approved by:

TITLE: **Control of Incoming Materials Used in Production**

AUTHOR: ______________________________

Name/Title/Department

____

Signature/Date

CHECKED BY: ______________________________

Name/Title/Department

____

Signature/Date

APPROVED BY: ______________________________

Name/Title/Department

____

Signature/Date

REVISIONS:

No.	Section	Pages	Initials/Date

SOP No. QCS-002.00 Effective date: mm/dd/yyyy
Approved by:

2.1 PURPOSE

To describe the control of incoming materials used in production.

2.2 RESPONSIBILITY

Quality control analysts are responsible to follow the procedure. The QC manager is responsible for the SOP compliance.

2.3 PROCEDURE

2.3.1 Storage Prior to Sampling

All coded raw and packaging materials received are kept in an expedition area to be sampled by QA before being transferred to the quarantine area. The sampling will be done in accordance with sampling instructions.

2.3.2 Overview of Sampling and Reporting Procedures

Inside the QC department the following procedures apply:

- All raw and packaging material samples are received at the sample receiving office in the QC department. The samples are verified against their respective documents, assigned a QC number, and sample details are entered in the respective sample registers as well through computer programs. These samples and documents are then distributed to various designated sections for further analyses.
- The samples are then tested in accordance with the approved specifications using specific test methods.
- The results are published in analytical reports designed for raw and packaging materials.
- The results are recorded in the raw material analysis journals maintained in chronological order. The same procedure applies for the packaging materials as well.
- If the material does not comply with the specifications, proper investigations are carried out as per the corresponding SOP, and after the investigation is completed, an analytical report is issued indicating the reason for the rejection of the lot.

2.4 REASON FOR REVISION

First time issued for your company, affiliates, and contract manufacturers.

QCS-003

YOUR COMPANY STANDARD OPERATING PROCEDURE

SOP No. QCS-003.00 Effective date: mm/dd/yyyy
Approved by:

TITLE: **Control of Bulk and Intermediate Products**

AUTHOR: ______________________________

Name/Title/Department

____

Signature/Date

CHECKED BY: ______________________________

Name/Title/Department

____

Signature/Date

APPROVED BY: ______________________________

Name/Title/Department

____

Signature/Date

REVISIONS:

No.	Section	Pages	Initials/Date

SOP No. QCS-003.00 Effective date: mm/dd/yyyy
Approved by:

3.1 PURPOSE

To describe the procedure for the handling of bulk and intermediate products in a QC laboratory.

3.2 RESPONSIBILITY

The process staff member of the QA department and laboratory analysts are responsible to follow the procedure. In turn, the QA and QC directors are responsible for the SOP compliance.

3.3 PROCEDURE

3.3.1 Bulk Products

Bulk products are sampled by the QA inspectors. The samples are then conveyed to QC laboratory along with the bulk release sheets (see Attachment I), also filled in by in-process quality assurance inspectors (QAI).

In the QC laboratory, the following procedures apply:

- The sample's QC number is recorded in the bulk product analysis journal, which is maintained in chronological order. The tests are carried out in accordance with specifications for finished products using analytical methods for finished products.
- If the sample complies with its specifications and the batch documents (raw material requisition, in-process control results, and batch manufacturing records) are evaluated and approved by the QC manager, then release for packaging is issued on a bulk release sheet which is later sent to the manufacturing department.
- The batch documents, processed so far, are filed. Additional batch documents from packaging operations are received after the packaging process is completed.
- If the sample does not comply with the specifications, a formal out-of-specification (OOS) investigation is carried out and accordingly the bulk release sheet is filled out indicating the reasons for the noncompliance of the batch and sent back to production to be processed further.

3.3.2 Intermediate Products

Intermediate products are sampled by the QA in-process inspector. The procedure carried out is similar to that in the case of bulk products. Release is effected by a release certificate issued by the QC manager.

3.4 REASON FOR REVISION

First time issued for your company, affiliates, and contract manufacturers.

3.5 ATTACHMENT FORM

ABC PHARMACEUTICAL COMPANY

LEVEL 4
SOP No.: QCS-003
Issued On:
Revision No.:
Initiator:

ATTACHMENT I

BULK RELEASE

Product name: ______________________ Batch No.: ___________ Code: ______
Manufacturing date: ___________ Expiry date: ___________

Manufacturing

- Ampoules	☐ Content	☐ Physical parameters		
- Capsules	☐ Dissolution	☐ Content (Assay)		
- Cream	☐ Content			
- Drops	☐ Content			
- Ointment	☐ Content			
- PPS	☐ Content			
- Suppositories	☐ Content	☐ Content uniformity		
- Suspensions	☐ Content	☐ Content uniformity		
- Syrups	☐ Content	☐ Physical parameters (pH, refractive index, etc.)		
- Tablets/caplets	☐ Granules	☐ Core	☐ Coated	☐ Content uniformity
- Vials	☐ Content			
- Others	☐ Content			

Remarks*____
____
____

Date: ________ Time: ______ Name & Signature: ____________________________

QC No.: ____ Date: _____ Time: ____ Signature: ____________________________

Physical Testing

____
____
____

Analyzed by: __________________________ Date: __________ Time: ___________

Chemical Testing

____
____
____

Analyzed by: __________________________ Date: __________ Time: ___________

☐ Released for packaging ☐ Not release for packaging

Checked by: QC officer **Approved by: QC manager**

* Other specific test needed

QCS-004

YOUR COMPANY STANDARD OPERATING PROCEDURE

SOP No. QCS-004.00 Effective date: mm/dd/yyyy
Approved by:

TITLE: **Checking of Finished Products**

AUTHOR: ______________________________

Name/Title/Department

____

Signature/Date

CHECKED BY: ______________________________

Name/Title/Department

____

Signaturc/Date

APPROVED BY: ______________________________

Name/Title/Department

____

Signature/Date

REVISIONS:

No.	Section	Pages	Initials/Date

SOP No. QCS-004.00 Effective date: mm/dd/yyyy
Approved by:

Subject: Checking of Finished Products

4.1 PURPOSE

To provide written instructions for checking of the finished products.

4.2 RESPONSIBILITY

This is the responsibility of all analysts testing the finished products to follow the procedure. The QC manager is responsible for the SOP compliance.

4.3 PROCEDURE

1. Full analysis for each and every batch will be done on bulk products (before packaging).
2. After full packaging of the batch (finished product), samples are collected from the packaging lines by the process control inspector, who will send samples with complete documents to the QC receiving room.
3. Samples of each consignee's quantity will be assigned a separate QC reference number.
4. Samples of products will be sent to finished and microbiological laboratories for checking according to the specification.
5. In the finished products laboratory, samples are checked for
   a. Description and appearance (for packaging)
   b. Packaging mode
   c. Content per pack
   d. Printed information (batch no., mfg./exp. dates, special printing as logo of consignee, agent name, price [where applicable], registration no., telephone No. [where applicable], etc.)
   e. Product description
   f. Filling volume or weight (of granules for suspension)
   g. Clarity, volume, specific gravity, and refractive index (for solution)
   h. Volume and viscosity (for suspension)
6. File documents sent by in-process control inspector will be verified against the accompanied checklist and this checklist will be signed by the finished product analysts.
7. Results of inspection are reported.
8. Microbiologically, the sample will be checked for total aerobic bacterial count, and for the absence of pathogens.
9. Reports and documents will be submitted to QC assistant manager, and in his/her absence to the physical laboratory in-charge for evaluation and finished product release.
10. If the finished product does not comply with specifications, the final decision about release or rejection is made by the QC manager/QA director or the Material Review Board.

4.4 REASON FOR REVISION

First time issued for your company, affiliates, and contract manufacturers.

QCS-005

YOUR COMPANY STANDARD OPERATING PROCEDURE

SOP No. QCS-005.00 Effective date: mm/dd/yyyy
Approved by:

Title: **Destruction of (Disposal) Samples after Complete Testing and Reference Samples after Expiry**

AUTHOR: ______________________________

Name/Title/Department

____

Signature/Date

CHECKED BY: ______________________________

Name/Title/Department

____

Signature/Date

APPROVED BY: ______________________________

Name/Title/Department

____

Signature/Date

REVISIONS:

No.	Section	Pages	Initials/Date

SOP No. QCS-005.00 Effective date: mm/dd/yyyy
Approved by:

5.1 PURPOSE

To provide written instructions to dispose or destroy the remaining samples after complete analysis and expiry of shelf life.

5.2 RESPONSIBILITY

This is the responsibility of the relevant section's supervisor (bulk or raw material) to follow the procedure. The QC manager will be responsible for system compliance.

5.3 PROCEDURE

Routine samples after complete testing:

1. After completing all analytical procedures, the samples remaining after analysis are treated as "pharmaceutical waste."
2. The liquid product will be discarded in a drainage system, while empty label bottles will also be treated as "pharmaceutical waste."
3. Dry (solid) products and the remaining packaging material will be put into plastic trash bags or carton boxes and labeled clearly as "pharmaceutical waste."
4. All pharmaceutical waste will be transferred to a pharmaceutical waste area.
5. General pharmaceutical waste disposal procedures will be followed. The waste is then transferred to the municipality area from where it is collected and dumped.
6. Until the waste is disposed or destroyed, it would be monitored by the cleaning personnel.

Reference samples after expiry:

7. A designated lab officer will check the data on a computer as well as inspect the reference sample room to determine whether the samples have completed 1 year after the expiry date.
8. The details of the expired file sample are entered in a log book (batch no., consignee, etc.).
9. Follow the same procedure from point No. 2.

5.4 REASON FOR REVISION

First time issued for your company, affiliates, and contract manufacturers.

QCS-006

YOUR COMPANY STANDARD OPERATING PROCEDURE

SOP No. QCS-006.00 Effective date: mm/dd/yyyy
Approved by:

TITLE: **Recording the Analysis and Controlling the Analytical Registers and Records**

AUTHOR: ______________________________

Name/Title/Department

____

Signature/Date

CHECKED BY: ______________________________

Name/Title/Department

____

Signature/Date

APPROVED BY: ______________________________

Name/Title/Department

____

Signature/Date

REVISIONS:

No.	Section	Pages	Initials/Date

SOP No. QCS-006.00 Effective date: mm/dd/yyyy
Approved by:

6.1 PURPOSE

To provide written instructions for recording the analysis and controlling analytical registers and records (journals, reports, and analysts log books).

6.2 RESPONSIBILITY

It is the responsibility of the sample receiving officer and analysts to follow up the procedure. The QC manager/director will be responsible for the SOP compliance.

6.3 PROCEDURE

1. Sample receiving registers
   - 1.1 Data on the received samples of raw and packaging material and the bulk and finished products are entered in their respective registers and the QC sample results are entered in the QC module by the sample receiving officer. After assigning a QC number the samples are distributed as per the procedure.
   - 1.2 The above-mentioned registers are used upto the last page and on completion they are labeled with starting and ending QC numbers including the respective dates and are archived.
2. Analysts registers
   - 2.1 Registers of raw and packaging materials as well as bulk and finished products analysis are maintained with each analyst of the respective section for entering the raw data of analysis.
   - 2.2 These registers are controlled by assigning a serial number for each one at the time of issuance.
   - 2.3 Completed registers are serially labeled by the user and returned to the corresponding section supervisor/incharge who then maintains them in an archive.
3. Analytical reports
   - 3.1 The raw material reports after final review by a supervisor and counter signed by a QC manager/director are filed in the corresponding records meant for raw material according to the code number.
   - 3.2 Bulk and finished product reports are filed with their respective batch files.
   - 3.3 Packaging material reports are filed in alphabetical order and stored in an archive.
4. Recording the analysis

   While recording the analysis, the analyst must consider and record the following:
   - a. Identification of the instrument
   - b. Preparation of diluent
   - c. Standards traceability
   - d. Lot number of working standard used
   - e. Mobile phases
   - f. Test solution
   - g. Recording of calculations pertaining to the test with representative units of measurement
   - h. Documentation of media used for microbiological testing

6.4 REASON FOR REVISION

First time issued for the ABC Pharmaceutical Company.

QCS-007

YOUR COMPANY STANDARD OPERATING PROCEDURE

SOP No. QCS-007.00 Effective date: mm/dd/yyyy
Approved by:

TITLE: **Testing of Finished Products**

AUTHOR: ______________________________

Name/Title/Department

____

Signature/Date

CHECKED BY: ______________________________

Name/Title/Department

____

Signature/Date

APPROVED BY: ______________________________

Name/Title/Department

____

Signature/Date

REVISIONS:

No.	Section	Pages	Initials/Date

SOP No. QCS-007.00 Effective date: mm/dd/yyyy
Approved by:

7.1 PURPOSE

To provide written instructions for checking the finished products.

7.2 RESPONSIBILITY

It is the responsibility of all analysts testing the finished product to follow up the procedure. The QC manager will be responsible for the SOP compliance.

7.3 PROCEDURE

1. A full analysis for each and every batch will be done on bulk product (before packaging).
2. After full packaging of the batch (the finished product), samples are collected from the packaging lines by the in-process control inspector, who will send samples with full documents to the QC receiving room.
3. Samples of each consignee's quantity will be given a separate QC reference number.
4. Samples of products will be sent to finished and microbiological laboratories for checking according to the specification.
5. In the finished products laboratory, samples are checked for
   a. Description and appearance (for packaging)
   b. Packaging mode
   c. Content per pack
   d. Printed information (batch no., mfg./exp. dates, special printing as logo of consignee, agent name, price (where applicable), registration no., telephone No. (where applicable), etc.)
   e. Product description
   f. Filling volume or weight (for PPS)
   g. Clarity, volume, specific gravity, and RI (for solution)
   h. Volume and viscosity (for suspension)
6. File documents sent by the in-process control inspector will be verified against the accompanied checklist and this checklist will be signed by the finished product analysts.
7. Results of inspection are reported.
8. Microbiologically, the sample will be checked for total aerobic bacterial count, and for the absence of pathogens.
9. Reports and documents will be submitted to the QC assistant manager, and in his/her absence to the physical laboratory in-charge for evaluation and release of the finished product.
10. If the finished product does not comply with specifications the final decision about release or rejection is made by the QC manager/QA director or the Material Review Board.

7.4 REASON FOR REVISION

First time issued for your company, affiliates, and contract manufacturers.

QCS-008

YOUR COMPANY STANDARD OPERATING PROCEDURE

SOP No. QCS-008.00 Effective date: mm/dd/yyyy
Approved by:

TITLE: **Microbiological Monitoring Program**

AUTHOR: ______________________________

Name/Title/Department

____

Signature/Date

CHECKED BY: ______________________________

Name/Title/Department

____

Signature/Date

APPROVED BY: ______________________________

Name/Title/Department

____

Signature/Date

REVISIONS:

No.	Section	Pages	Initials/Date

SOP No. QCS-008.00 Effective date: mm/dd/yyyy
Approved by:

8.1 PURPOSE

This SOP describes the microbiological monitoring program at ABC Pharmaceutical Company. The SOP covers sampling, testing, and interpretation of results in the final report. The objective of monitoring program is to provide sufficient information to control the microbiological quality of water for its intended use.

8.2 SCOPE

This SOP is intended for microbiological monitoring of water systems used in the QC laboratory as well as in the Research and Development (R&D) laboratory.

8.3 RESPONSIBILITY

8.3.1 QC Department

8.3.1.1 Microbiologist

QC microbiologist will carry out the following activities:

- Prepare the testing equipment and growth media and carry out the growth promotion test on the used media
- Assure that testing equipment, utilities, and media are sterile and in appropriate condition during testing
- Follow up the sampling schedule
- Assure that sampling techniques are applied appropriately
- Count and identify the microbial isolates
- Document in the final report, the sampling techniques used and all related details including results

8.3.1.2 Supervisor

QC microbiology lab supervisor will

- Check and interpret the results
- Determine and report the out-of-specification results
- Summarize the data

8.3.1.3 QC/QA Director

- Evaluate and approve the results
- Review the trend analysis
- Review the cause of out-of-specification results
- Deviations from SOP should be noted and approved by QC and QA managers

8.3.2 R&D Laboratory

The R&D laboratory will validate the applied methodology as per the approved protocol.

SOP No. QCS-008.00 Effective date: mm/dd/yyyy
Approved by:

8.3.3 QA Department

8.3.3.1 QA Manager

The QA manager will review the data and will also follow them up with the corrective actions recommended in the reports.

8.3.3.2 QA Director

QA director will recommend the remedial actions for OOS results.

8.3.4 Training

A formal personnel training program is required to minimize the risk of false-positive or false-negative results.

The training should include at least

- Instructions on the basic principles of aseptic technique
- Good sampling practices
- Demonstration of water system and its potential sources of contamination
- Instructions on the basic principles of microbiology, microbial physiology, disinfection, sanitation, media selection, media preparation, taxonomy, and sterilization as required by the nature of personnel involvement in aseptic processing
- Endotoxin basics, sources, and determination

8.3.4.1 Supervisor

- Personnel involved in microbial identification will require specialized training on required laboratory methods.
- Additional training on the management of the water data collected.
- Knowledge and understanding of applicable operating procedures, especially those relating to corrective measures that are taken when an adverse drift in microbiological conditions are detected.
- Understanding of regulatory compliance policies and each individual's responsibilities with respect to good manufacturing practices (GMP) should be an integral part of the training program as well as training in conducting investigations and in analyzing data.

8.3.5 Sampling

8.3.5.1 Preparation of the Sampling Bottles

Wash the bottles carefully with water for injection (WFI)

- *WFI sampling bottles:* The samples are collected in bottles with caps depyrogenated at 250°C/30 min.
- *Purified water sampling bottles:* The samples are collected in autoclavable wide-mouthed bottles with stoppers. Glass or polypropylene bottles may be used. Sterilize the bottle at 180°C/1 h (dry heat) or autoclave at 121°C/15 min.

8.3.5.2 Sample Size

The size of the samples should be sufficient for all required tests, but not less than 100 mL/sample.

SOP No. QCS-008.00 Effective date: mm/dd/yyyy
Approved by:

8.3.5.3 Sampling Technique

8.3.5.3.1 Special Precautions to Be Taken during Sampling

- Follow aseptic technique rules. Wear proper garments, sterile gloves, and mask.
- Keep the sampling bottle away and do not open it until the sampling point is ready.
- Be sure that the sampling point is in proper condition and under high pressure.
- Avoid leaky fixtures that may contaminate the samples with bacteria from the outside surface.
- Disinfect the sampling point from inside and its surrounding with 70% ethyl alcohol or 3% H_2O_2 for 2 min.

8.3.5.3.2 Special Precautions during Sampling

- *Pure steam:* Through an autoclavable condenser, pass the steam for 30 min and then collect the sample in a tightly closed container with a vent passage.
- *Bacterial endotoxin:* Through apyrogenic connection, pass the water for about 1 min and then collect the sample in apyrogenic bottle. All materials coming in contact with test materials and reagents must be pyrogen free. Careful techniques must be used to prevent contamination with environmental endotoxin.

8.3.5.3.3 Multistep Sampling Methods

- Open the sampling point to maximum for about 30 s with pulse flushing (quick open and close) to purge any dust and residual of the disinfectant.
- Remove the cap from the bottle immediately prior to obtaining a sample.
- The cap may be placed on a clean surface with the open side facing upwards. The sample side of the cap should not come in contact with any surface, including the fingers or the hand of the individual during sampling.
- Adjust the sampling point to an adequate level, open the sampling bottle under a stream of water in a downward direction and then invert the bottle quickly upward to collect the sample. Fill the bottle, without letting the sample to overflow. Allow sufficient space (1″ or 2.5 cm) at the top of the bottle to facilitate mixing of the sample by shaking.
- Do not allow the bottle or the water in the bottle to come in contact with the source.
- Remove the bottle from the sample stream and place the cap on it as quickly as possible.
- Samples must be properly labeled, and must include the following:

Date:		Name of sampler:	
Time of sampling:		Sampling point:	
Special notes:			

Be sure that the stopper of the bottle is not mishandled or in contact with any source of contamination. Cover the neck of the bottle with a new sterile wrapped material.

8.3.5.3.4 MicropreSure® In-Line Filtration Sampler

- Preparation of in-process water samples using the MicropreSure® device.

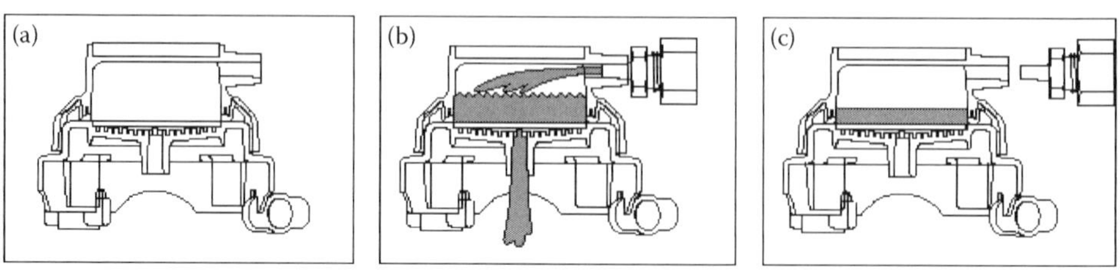

SOP No. QCS-008.00 Effective date: mm/dd/yyyy
Approved by:

- Flush the sanitary sampling valve port thoroughly, and then close the valve.
- Remove the yellow and blue stoppers and attach them to the base.
- Fit the MicropreSure outlet to the receiving graduated container, or connect a length of tubing to the device outlet.

- Insert the sampling valve port outlet into the MicropreSure inlet and, with a slight twisting movement, push the MicropreSure device onto the valve.
- Close the sampling valve when the desired sample volume or the desired sampling time has been reached.
- Gently pull the MicropreSure device away from the valve, using a rotating movement. Maintain the device in a horizontal position.
- Place the yellow stopper over the MicropreSure device outlet and use the blue stopper to seal the device inlet.
- Record the sample location, time, and volume on the side of the MicropreSure dome and send to the laboratory for processing.

- Access the membrane by connecting the MSOpener™ to a vacuum source, and placing the MicropreSure In-Line Filtration Sampler on top of the MSOpener.

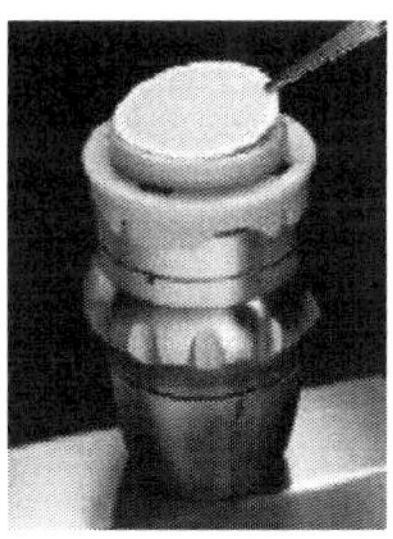

- Incubate the membrane by lifting the membrane and transferring it onto a media plate.

SOP No. QCS-008.00 Effective date: mm/dd/yyyy
Approved by:

8.3.5.4 Sampling Points of Sterile Plant WFI (Attachment I)

Building	Sampling Point Description	Activity	Frequency[a]
Sterile plant		WFI return	Each manufacturing day or at least twice a week
		WFI supply	
		WFI return from freeze drier	
		WFI supply for freeze drier	
		Solution preparation	
		Washing machine	
		Vessels machine (sink)	
		Clothes washing	Twice a week

[a] It is appropriate to increase or decrease sampling based on the trend performance. The time and date can be determined according to the activity.

8.3.5.5 Pure Steam (Attachment II)

Building	Sampling Point Description	Activity	Frequency[a]
Sterile plant		Station	Once a week
		SIP	
		SIP	

[a] It is appropriate to increase or decrease sampling based on the trend performance. The time and date can be determined according to the activity.

8.3.5.6 Sampling Points of Biotechnology Plant

- The ultrapurification water system can be sampled each manufacturing day or at least once weekly.
- Deionized water for washing can be sampled each manufacturing day or at least once weekly.

8.3.5.7 Sampling Points of Sterile Plant (DI Water)

Building	Sampling Point Description	Activity	Frequency[a]
Sterile plant A		DI return	Once a week
		DI supply	
		Hand washing	Twice a week
		Washing	
		Solution preparation	
		Washing machine	
		Sink	

SOP No. QCS-008.00 Effective date: mm/dd/yyyy
Approved by:

(continued)

Building	Sampling Point Description	Activity	Frequency[a]
Sterile plant B		DI return	Once a week
		DI supply	
		Tools washing	
		Clothes washing	
		Utensils washing	
		Media preparation	
		Washing	
Sterile plant C		DI return	Once a week
		DI supply	
		Tools washing	
		Clothes washing	
		Utensils washing	
		Media preparation	
		Washing	

[a] It is appropriate to increase or decrease sampling based on the trend performance. The time and date can be determined according to the activity.

8.3.5.8 Sampling Points of a Nonsterile Plant (Solid Dosage)

Room No.	Sampling Points Description	Activity	Type	Frequency[a]
		Solution preparation	Hot Cold	Once weekly
		Washing small parts	Cold	Once weekly
		Granulation & drying-I	Hot Hot Cold	Once weekly
		Granulation & drying-II	Hot	Once weekly
		Tablet compression-II	Hot Cold	Once weekly
		Tablet compression-III	Hot Cold	Once weekly
		Tablet compression-IV	Hot Cold	Once weekly

continued

SOP No. QCS-008.00 Effective date: mm/dd/yyyy
Approved by:

(continued)

Room No.	Sampling Points Description	Activity	Type	Frequency[a]
		Tablet compression-V	Hot Cold	Once weekly
		Sugar coating	Hot Cold	Once weekly
		Technical room-I	Hot Cold	Once weekly
		Film coating-I	Hot	Once weekly
		Film coating-II	Hot	Once weekly
		Film coating-III	Hot	Once weekly
		Washing area	Hot Hot Cold Cold	Once weekly

[a] It is appropriate to increase or decrease sampling based on the trend performance. The time and date can be determined according to the activity.

Note: The sampling points are not tested due to unavailability of access. However, all finished products are tested.

8.3.5.9 Sampling Points of a Nonsterile Plant (Liquid Dosage)

Room No.	Sampling Points Description	Activity	Type	Frequency[a]
		Raw material storage tanks	Hot Hot Cold Cold	Twice a week
		Solution preparation room	Hot Cold	
		Washing area	Hot Cold	
		Semisolid manufacturing room	Hot Hot	

[a] It is appropriate to increase or decrease sampling based on the trend performance. The time and date can be determined according to the activity.

SOP No. QCS-008.00 Effective date: mm/dd/yyyy
Approved by:

8.3.5.10 Sampling Points of R&D Laboratory

Room No.	Sampling Point Description	Type	Frequency[a]
		Cold	Once a month
		Cold	Once a month
		Cold	Once a month
		Cold	Once a month
		Cold	Once a month
		Cold	Once a month
		Cold	Once a month
		Cold	Once a month
		Cold	Once a month
		Cold	Once a month
		Cold	Once a month
QC washing room		Cold	Once a month
		Cold	Once a month
		Cold	Once a month

[a] It is appropriate to increase or decrease sampling based on the trend performance. The time and date can be determined according to the activity.

8.3.5.11 Sampling Points of QC Laboratory

Room No.	Activity	Type	Frequency[a]
	Microbiology lab	Cold	Once weekly
	Chemical lab	Cold	Once weekly

[a] It is appropriate to increase or decrease sampling based on the trend performance. The time and date can be determined according to the activity.

8.3.5.12 Feed Water and Drinking Water

Building	Sampling Point Description	Activity	Frequency[a]
		Feed water	Once monthly
		R.O.	Once monthly
		Deionizer	Once monthly
		After filtration	Once monthly

continued

SOP No. QCS-008.00 Effective date: mm/dd/yyyy
Approved by:

(continued)

Building	Sampling Point Description	Activity	Frequency[a]
		Chilled	Once monthly
		Feed water	Once monthly
		R.O.	Once monthly
		Deionizer	Once monthly
		After filtration	Once monthly
		Chilled	Once monthly

[a] It is appropriate to increase or decrease sampling based on the performance. The time and date can be determined according to the activity.

8.3.5.13 Sampling Points of Penicillin Plant (WFI)

Sampling Point Description	Activity	Frequency[a]
	WFI tank	Each operating day or at least twice a week
	Supply from tank	
	Return from loop	
	Washing machine	
	Washing clothes	
	Tools washing	
	QC lab (sterility room)	

[a] It is appropriate to increase or decrease sampling based on the trend performance. The time and date can be determined according to the activity.

8.3.5.14 Sampling Points of Penicillin Plant (DI Water)

Sampling Point Description	Activity	Frequency[a]
	Supply from tank	Once a week
	Return from loop	Once a week
	QC lab	Once a week
	QA lab	Once a week
	Storage tank	Once a week
	Steam generator	Once a week

[a] It is appropriate to increase or decrease sampling based on the trend performance. The time and date can be determined according to the activity.

SOP No. QCS-008.00 Effective date: mm/dd/yyyy
Approved by:

8.3.5.15 Sampling Points of Cephalosporin Plant (WFI)

Sampling Point Description	Activity	Frequency[a]
	Supply from general tank	Each operating day or at least twice a week
	Return from general loop	
	WFI tank	
	Supply from tank	
	Return from loop	
	Washing machine	
	Washing clothes	
	Tools washing	
	QC lab (sterility room)	

[a] It is appropriate to increase or decrease sampling based on the trend performance. The time and date can be determined according to the activity.

8.3.5.16 Sampling Points of Cephalosporin Plant (DI Water)

Sampling Point Description	Activity	Frequency[a]
	Supply from tank	Once a week
	Return from loop	Once a week
	QC lab	Once a week
	QA lab	Once a week
	Storage tank	Once a week
	Steam generator	Once a week
	For distillator	Once a week

[a] It is appropriate to increase or decrease sampling based on the trend performance. The time and date can be determined according to the activity.

8.3.5.17 Pure Steam Penicillin and Cephalosporin Plants

Building	Sampling Point	Activity	Frequency[a]
Penicillin plant	Pure steam	Autoclave	Once a week
Cephalosporin plant	Pure steam	Autoclave	

[a] It is appropriate to increase or decrease sampling based on the trend performance. The time and date can be determined according to the activity.

SOP No. QCS-008.00 Effective date: mm/dd/yyyy
Approved by:

8.3.5.18 Water Treatment Penicillin and Cephalosporin Plants

Sampling Point Description	Activity	Frequency[a]
	Feed water	Once/3 months
	After sand filter I	Once/3 months
	After sand filter II	Once/3 months
	After 5 μ cartridge	Once/3 months
	After RO	Once/3 months
	RO tank	Once/3 months
	After deionizer	Once monthly
	After UV lamp	Once monthly
	Supply water	Once monthly
	Return water	Once monthly

[a] It is appropriate to increase or decrease sampling based on the performance. The time and date can be determined according to the activity.

8.3.6 General Instructions for Water Sample Manipulation

- Deliver the sample to the laboratory as fast as possible, and the maximum time is 2 h after that the sample will be unacceptable to microbiological analysis.
- Maintain the quality of the sample through the transport to the laboratory, especially the neck of the bottle must not be mishandled.

8.3.7 Micro Count Methodology Consideration

- The method selected must be capable of isolating the numbers and types of organisms that have been estimated significant relative to the system control and product impact for each individual system. The recommended method is the membrane filtration method, whereas pour plate and the most probable number may be used as per requirements.
- Culture approaches for each type of water are further defined by the type of medium used in combination that should be selected according to the monitoring needs presented by a specific water system as well as its ability to recover microorganisms that could have a detrimental effect on the product or the process.

Cultivation on low-nutrient media and incubation at 20–25°C for 5–7 days (recommended). The applied approach can be documented and validated.

1. *Media growth promotion:* Media are able to support growth when inoculated with less than 100 colony-forming units (CFU) of the challenge organisms as per STM MC 000. The media are capable of supporting growth of indicator microorganisms and of environmental isolates

SOP No. QCS-008.00 Effective date: mm/dd/yyyy
Approved by:

from samples obtained through the monitoring program or their corresponding ATCC (American Type Culture Collection) strains. The ability of the selected media to detect and quantitate these anaerobes or microaerophilic microorganisms can be evaluated.

2. *Controls:* During routine monitoring, each preparation of media can be tested for positive control (with 100 CFU) and negative control (open during performing the test).
3. *Methodology* "according to the type of water."
   a. Pure steam or water for injection.
      i. *Total aerobic viable count:* Using the membrane filtration method, a 100 mL sample can be filtrated and placed on low-nutrient-content media at 20–22°C for 5 days and observed for another 2 days.
      ii. *Bacterial endotoxin:* Kinetic test or gel clot limit method "STM MC 000."
   b. *Purified water:* There are three types of purified water.
      i. *Class A purified water:* It is used for highly bacteria critical products such as certain topical, antacids, and inhalants as well as the deionized water itself.
         A. *Total aerobic viable count:* As per STM MC 001 using the membrane filtration method. A sample of 100 mL can be filtrated and placed on low-nutrient media, at 20–22°C for 5 days and observed for another 2 days.
         B. *Pseudomonas:* A sample of 100 mL can be filtrated and the membrane filter is placed in 50 mL of TSB (tryptic soy broth) incubated for 20–28 h at 35 + 2°C. The membrane filter is then transferred into Pseudomonas agar F and P for 40–48 h at 35 + 2°C. If any characteristic colony is detected, identify using API and/or automated identification system.
         C. *Coliform:* A sample of 100 mL can be filtrated and, the membrane filter is placed in 50 mL of MacConkey broth, incubated for 20–28 h at 35 + 2°C. Later it is transferred into EMB agar for 40–48 h at 35 + 2°C. If any characteristic colony is detected, identify using the API and/or automated identification system.
      ii. *Class B purified water used for rinsing:* Same as purified water class A except for microbial limit is 100 CFU/100 mL.
      iii. *Class C purified water:* It has minimum concerns associated with bacteria levels such as washing water and laboratory usage.
         A. *Total aerobic viable count:* Pour 1 mL into a plate containing 25 mL of tryptone glucose extract agar or TSA 1/2 conc. or 1/2 conc. plate count agar. Incubate at 22°C/day.
         B. *Pseudomonas:* Pour 1 mL into a plate containing 25 mL of Pseudomonas agar F or P. Incubate at 35–37°C/2 days.
   c. *Feed water, RO water, and drinking water:* The drinking water depends on the local availability and is delivered by the municipal or other local public system or drawn from a private well or reservoir. This water serves as the starting material for most forms of water.
      i. *Total aerobic viable count:* Pour 1 mL into a plate containing 25 mL of TSA 1/2conc. or 1/2conc. plate count agar. Incubate at 22°C/5 days.
      ii. *Pseudomonas:* Pour 1 mL into a plate containing 25 mL of Pseudomonas agar F or P. Incubate at 35–37°C/2 days.
      iii. *Coliform:* Using the membrane filtration technique, filtrate 100 mL and place the membrane filter into Endo agar and identify using the API and/or automated identification system.

iv. *Faecal streptococci:* Using direct inoculation of 1 mL into TSB, if there is turbidity, transfer into azide dextrose broth or azide blood agar and identify using the API and/or automated identification system.

8.3.8 General Instruction for Water Analysis

Analyze the sample as fast as possible. In case of a probable delay, keep it in a refrigerator for a maximum of 2 h. Never incubate or keep the sample at room temperature.

An appropriate level of control may be maintained by using data trending techniques and limiting specific contraindicated microorganisms.

8.3.9 Identification of Microbial Isolates

- For general purposes: Colony morphology, gram staining, or microscopic morphology as a routine may be sufficient.
- Pheno-typical isolates from purified water can be characterized. Biochemical testing, such as oxidase test, urease test, catalase test, citrate test, coagulase test, and indole test and commercial test kits such as API tests and reagents may be used for conformation of some unique isolates. Identification of isolates from critical sampling points such as water used for preparation and areas immediate to these critical sampling points such as water used for clean in place (CIP) should take precedence over identification of microorganisms from noncritical sampling points.
- Isolates from WFI can be identified to species level (if possible) using API tests/automated identification system. This pertains to the detection of any isolate obtained from a sample that breaches either the alert or action level.
- Mold does not carry "special" status relative to bacteria. Any significant shifts in type or number require action—regardless of mold vs. bacteria.

  Note: The organism recovered from production environments may be highly stressed due to physical factors, contact with chemicals, and thermal stress. It may be difficult to obtain typical biochemical reactions with these isolates. The databases for commercial test kits and ID systems are often designed for clinical isolates and may be incomplete with regard to industrial isolates. Thus, interpretation of such microbial data requires experienced judgment.

- Identification methods should be verified, and ready-to-use kits should be made available for their intended purpose. The methods used for identification of isolates should be verified using indicator microorganisms.
- The information gathered by an identification program can also be useful in the investigation of the source of contamination, especially when the action levels are exceeded.

8.3.10 Result

1. *Result validity:* The result can be invalid if
   a. Either negative control or positive control was unacceptable
   b. The used media or reagent is unacceptable
   c. Unusual events during sampling
   d. The sample subjected to factor changing its original content such as extent transportation time, incubate at high temperature or improper handling
      i. Carry out a laboratory technical investigation and retest the testing reagents
      ii. Repeat the test with more restrict sampling and testing conditions

SOP No. QCS-008.00 Effective date: mm/dd/yyyy
Approved by:

iii. Carry out investigational analysis of OOS microbiological results
iv. If the new sample is OOS the sampling point is rejected

2. *Trend analysis:* Trend—periodic printouts or tabulations of results for their purified water systems. These printouts or data summaries should be reviewed. A trend analysis is used to facilitate decision-making for requalification of a controlled environment or for maintenance and sanitization schedules.

   Interpretation of the significance of fluctuations in counts or a change in flora should be based on the experienced judgment of qualified personnel.
3. *Limits:* Limits are conservative measures designed to signal potential or actual drift from historical or design performance characteristics. They are not extensions of product specifications, but are intended to flag changes so that corrective action may be taken before product quality is adversely affected.

Not all situations require use of both alert and action limits.

4. *Alert levels:* Alert levels are specific for a given facility and are established on the basis of a baseline developed under an environmental monitoring program. These levels are usually reexamined for appropriateness at an established frequency. When the historical data demonstrate improved conditions, these levels can be reexamined and changed to reflect the conditions.

Exceeding the alert level is not essential for definitive corrective action, but it should at least prompt a follow-up investigation that could include sampling plan modifications.

5. *Action levels:* An action level in microbiological environmental monitoring is that when the level of microorganisms exceeds requires immediate follow-up and, if necessary, corrective action. The evaluation does not depend on the number of colonies only but also on the types of microbes isolated and the suspected hazard that may result.

Regarding microbiological results, for pure steam and water for injection, it is expected that they be essentially sterile. Since sampling is frequently carried out in nonsterile areas and is not truly aseptic, occasional low-level counts due to sampling errors may occur. Agency policy is that less than 5 CFU/100 mL is an acceptable action limit.

Type of Water	Alert Limit	Action Limit
Pure steam	≤1 CFU/100 mL *Endotoxin*: LT 0.06 USP EU/mL	≤5 CFU/100 mL *Endotoxin*: LT 0.125 USP EU/mL
Water for injection	≤3 CFU/100 mL *Endotoxin*: LT 0.125 USP EU/mL	≤5 CFU/100 mL *Endotoxin*: LT 0.25 USP EU/mL
Ultra pure water	≤5 CFU/100 mL *Endotoxin*: LT 0.125 USP EU/mL	≤10 CFU/100 mL *Endotoxin*: LT 0.25 USP EU/mL
Purified water *used highly bacteria critical products*	≤10 CFU/100 mL	≤20 CFU/100 mL *Pseudomonas*: absence in 100 mL *Coliform*: absence in 100 mL
Purified water used for rinsing	≤50 CFU/100 mL	≤100 CFU/100 mL *Pseudomonas*: absence in 100 mL *Coliform*: absence in 100 mL

continued

SOP No. QCS-008.00 Effective date: mm/dd/yyyy
Approved by:

(continued)

Type of Water	Alert Limit	Action Limit
Purified water used for washing and lab usage	≤25 CFU/mL	≤100 CFU/mL *Pseudomonas*: absence in 1 mL
R.O. water Feed water Drinking water	NA	*Total viable count*: ≤500 CFU/mL *Coliform*: absence in 100 mL *Pseudomonas*: absence in 1 mL *Faecal streptococci*: absence in 1 mL

The data recovered from the sampling is reviewed for conformance to specifications. Maintenance and production management are notified immediately when action levels are exceeded. Results which are over action level (OALs) cause initiation of an investigation to identify the source of the contaminant. Bacteria which are recovered are identified to genus whenever possible. Sampling may be carried out more frequently during the investigation or until counts fall back within limits. System sanitization will also be carried out. The specifications for monitoring of water for injection are defined in manufacturing site SOP.

8.3.11 Interpretation of Data

Since microbiological test results from a water system are not usually obtained until after the drug product is manufactured, results exceeding limits can be reviewed with regard to the drug product formulated from such water. Consideration with regard to the further processing or release of such a product will be dependent upon the specific contaminant, the process, and the end use of the product. Such situations are usually evaluated on a case-by-case basis.

8.3.11.1 Corrective Action

As per the attached flowchart (Attachments A and B for excursions of WFI and DI water): When action limits are exceeded the QA manager will investigate the cause of the problem, take action to correct the problem, and assess the impact of the microbial contamination on products manufactured with the water and document the results of their investigation.

In addition to reviewing test results, summary data, investigation reports, and other data, the print of the system should be reviewed when conducting the actual physical inspection. As pointed out, an accurate description and print of the system is needed in order to demonstrate that the system is validated.

Maintenance and production management are notified immediately when action levels are exceeded. Results which are OALs cause initiation of an investigation to identify the source of the contaminant. Bacteria which are recovered are identified to the genus level whenever possible. Sampling may be carried out more frequently during the investigation or until counts fall back within limits. System sanitization will also be carried out.

8.4 REASON FOR REVISION

First time issued for the ABC Pharmaceutical Company.

8.5 ATTACHMENT FORMS

Attachment 1A
SOP No.:
Issued on:
Revision No.:
Initiator name:

ABC PHARMACEUTICAL COMPANY
QC LABORATORY

MICROBIOLOGICAL MONITORING
WFI TESTING REPORT

Sampled from: **Testing date:**
Sampling time: **Sampled by:**

Sampling Point	TMC	LAL
Return WFI	CFU/100 mL	USP EU/mL
Supply WFI	CFU/100 mL	USP EU/mL
xxx	CFU/100 mL	USP EU/mL
xxx	CFU/100 mL	USP EU/mL
xxx	CFU/100 mL	USP EU/mL
xxx	CFU/100 mL	USP EU/mL

Action limit: TMC : 5 CFU/100 mL LAL : NMT 0.25 USP EU/mL

Comment :

Tested by : **Report Date:**

Checked by : Manager QC

QC director

CC: QA Production Maintenance director
In case of failure Manufacturing director
QA director

Attachment II

ABC PHARMACEUTICAL COMPANY
QC LABORATORY

MICROBIOLOGICAL MONITORING
STEAM TESTING REPORT

Sampled from: Sterile plant **Testing date:**...................

Sampling time: .... **Sampled by:**....................

Sampling Point	Total Microbial Count	LAL
Steam	CFU/100 mL	USP EU/mL
xxx	CFU/100 mL	USP EU/mL
xxx	CFU/100 mL	USP EU/mL
Comment :	....	

Action Limits: TMC: 5 CFU/100 mL LAL: NMT 0.125 USP EU/mL

Tested by: **Report date:**

Checked by: QC manager

QC director

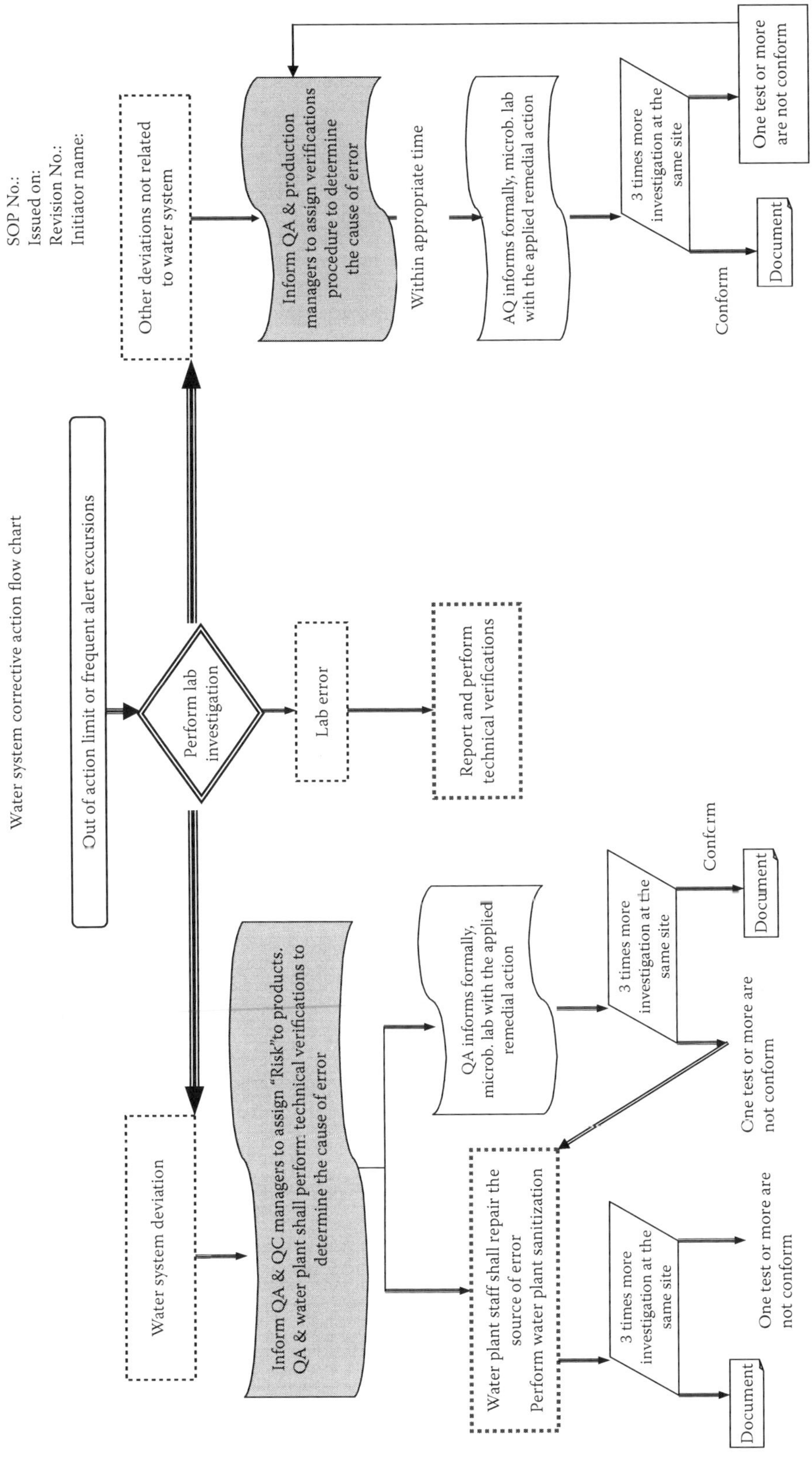

Water system corrective action flow chart
SOP No.:
Issued on:
Revision No.:
Initiator name:
Out of action limit or frequent alert excursions
Perform lab investigation
Lab error
Report and perform technical verifications
Water system deviation
Inform QA & QC managers to assign "Risk"to products. QA & water plant shall perform technical verifications to determine the cause of error
QA informs formally, microb. lab with the applied remedial action
3 times more investigation at the same site
Conform
Document
One test or more are not conform
Water plant staff shall repair the source of error Perform water plant sanitization
3 times more investigation at the same site
One test or more are not conform
Document
Other deviations not related to water system
Inform QA & production managers to assign verifications procedure to determine the cause of error
Within appropriate time
AQ informs formally, microb. lab with the applied remedial action
3 times more investigation at the same site
Conform
Document
One test or more are not conform

ABC PHARMACEUTICAL COMPANY

LEVEL 3
PM FORM
SOP No.:
Issued on:
Revision No.:
Initiator name:

WATER BIOBURDEN EXCURSION INVESTIGATION REPORT

Attachment A

Date:.......................... Time: .... Sampling point: ....
Used for:.. Batch No.: ....

ALERT LIMIT	ACTION LIMIT	RESULTS

Trend: Isolated excursion ☐ Frequent excursion ☐
Attached the trend report of the point for the last 3 months
Comment upon the point during sampling: ....
Microbial identification: ....
....

Manager QC

QA INVESTIGATION

Carried out by: **Reviewed by:**

Water system temp. trends (if any): ....
Chemical trend reports: ....
Feed water pretreatment trend reports: ....
Unusual events during sampling day: ....
Other environmental monitoring data in the area: ....
Water system sanitization records: ....
Status of water pumps: ....
Status of main tank: ....
Status of vent filter: ....
Status of automatic valves: ....
Status of manual valves: ....
Status of sampling valves: ....
Status of sampling connection (if any): ....
CIP of preparation tank: ....CIP of holding tank: ....
Status of manual valves: ....

Comment: ....
....

Corrective actions **(Carried out by utilities engineer):** ....
....

Conclusion: ....
....

QC director QA director

CC: QA file/utilities engineer/production manager/QC "microbiology"
In case of NC investigation report to be raised by QAI

ABC PHARMACEUTICAL COMPANY

LEVEL 3
PM FORM
SOP No.: QCS-008
Issued on:
Revision No.:
Initiator name:
S. No.: ....

ULTRAPURE WATER BIOBURDEN EXCURSION INVESTIGATION REPORT

Attachment B

Date:........................... Time: .... Sampling point:..............................
Used for:... Batch No.:......................................

ALERT LIMIT	ACTION LIMIT	RESULTS

Trend: Isolated excursion ☐ Frequent excursion ☐
Attached the trend report of the point for the last 3 months
Comment upon the point during sampling: ....
Microbial identification: ....
....

Manager QC

QA INVESTIGATION

Carried out by: **Reviewed by:**

Ultrapure system temp. trends (if any): ....
Review of the operational parameters: ....

Chemical trend reports: ....
Feed water pretreatment trend reports: ....
Unusual events during sampling day: ....
Other environmental monitoring data in the area: ....
Water system sanitization records: ....
Status of storage tank: ....
Status of vent filter: ....
Status of sampling valves: ....
Status of sampling connection (if any): ....

Comment: ....
....

Corrective actions **(Carried out by utilities engineer):** ....
....

Conclusion: ....
....

QC director QA director

CC: QA file/utilities engineer/production manager/QC "microbiology"

In case of NC investigation report to be raised by QAI

Attachment III
SOP No.:
Issued on:
Revision No.:
Initiator name:

ABC PHARMACEUTICAL COMPANY
QC LABORATORY

MICROBIOLOGICAL MONITORING
PURIFIED TESTING REPORT

Sampled from: Solid dosage from plant **Testing date:** ....

Sampling time: .... **Sampled by:** ....

Sampling Point	TMC	Coliform	Pseudomonas
Film coating I	CFU/100 mL		
Sugar coating hot	CFU/100 mL		
Sugar coating cold	CFU/100 mL		
Sample preparation hot	CFU/100 mL		
Sample preparation cold	CFU/100 mL		
Washing I hot	CFU/100 mL		
Washing I cold	CFU/100 mL		
Basement I hot	CFU/100 mL		

Action limit: TMC: 100 CFU/100 mL, Pseudomonas: Absent/100 mL, Coliform: Absent/100 mL

Comment:

Tested by: **Report Date:**

Checked by: QC manager

QC director

CC: QA manager, production manager, water treatment in-charge
In case of failure QA director, manufacturing director, maintenance director

Attachment IV
SOP No.:
Issued on:
Revision No.:
Initiator name:

ABC PHARMACEUTICAL COMPANY
QC LABORATORY

MICROBIOLOGICAL MONITORING
PURIFIED WATER TESTING REPORT

Sampled from: Semisolid dosage plant **Testing date:**...

Sampling time: .... **Sampled by:** ....

Room No.	Sampling Point	Total Aerobic Count	Coliform	Pseudomonas
SILOS	Outlet		CFU/100 mL	
	Inlet		CFU/100 mL	
Preparation	01 VALVE 1		CFU/100 mL	
	01 VALVE 2		CFU/100 mL	
Washing	01 VALVE 1		CFU/100 mL	
	01 VALVE 2		CFU/100 mL	
Semisolid Manufacturing	Sample 1		CFU/100 mL	
	Sample 2		CFU/100 mL	
Comment:				

Action Limits: TAC: 20 CFU/100 mL, Pseudomonas & Coliform: Absent in 100 mL

Tested by: **Report Date:**

Checked by: QC manager

QC director

Attachment V
SOP No.:
Issued on:
Revision No.:
Initiator name:

ABC PHARMACEUTICAL COMPANY
QC LABORATORY

MICROBIOLOGICAL MONITORING
PURIFIED WATER TESTING REPORT

Sampled from: R&D laboratory **Testing date:** ....

Sampling time: .... **Sampled by:** ....

Room No.	Sampling Points	Total Aerobic Count	Pseudomonas
245			CFU/mL
244			CFU/mL
243			CFU/mL
P.05			CFU/mL
242			CFU/mL
QC washing room			CFU/mL
			CFU/mL
			CFU/mL
229			CFU/mL
230			CFU/mL
231			CFU/mL
232			CFU/mL
			CFU/mL
			CFU/mL
Comment:	....		

Action limits: TAC: 100 CFU/mL, Pseudomonas: Absent in 1 mL

Tested by: **Report Date:**

Checked by: QC manager

QC director

QCS-009

YOUR COMPANY STANDARD OPERATING PROCEDURE

SOP No. QCS-009.00 Effective date: mm/dd/yyyy
Approved by:

TITLE: **Clothing in Microbiology Laboratory**

AUTHOR: ______________________________

Name/Title/Department

____

Signature/Date

CHECKED BY: ______________________________

Name/Title/Department

____

Signature/Date

APPROVED BY: ______________________________

Name/Title/Department

____

Signature/Date

REVISIONS:

No.	Section	Pages	Initials/Date

SOP No. QCS-009.00 Effective date: mm/dd/yyyy
Approved by:

Subject: Clothing in Microbiology Laboratory

9.1 PURPOSE

To provide written instruction for clothing in microbiology laboratories.

9.2 RESPONSIBILITY

It is the responsibility of the QC analyst (or microbiologist) to follow up the procedure. The QC manager will be responsible for the SOP compliance.

9.3 SAFETY PRECAUTION

Minimize your visits to other sections of the department to avoid contamination.

9.4 PROCEDURE

1. Wear a clean white coat and white clogs before entering the microbiology lab. The coats will be available in a poly bag marked "Washed Coats," lying in a cupboard near the QC entrance.
2. Prior to the aseptic technique wear a lint-free coat, a mask, a disposable head cover, a pair of sterile disposable gloves, and a face mask.
3. At the end of the day, deposit the lab coats in a poly bag marked "Used Coats." All used lab coats will be lying in another cupboard at the entrance of the QC department.
4. Everyday the laboratory cleaner will transport the used lab coats to the laundry for cleaning.

9.5 REASON FOR REVISION

First time issued for the ABC Pharmaceutical Company.

QCS-010

YOUR COMPANY STANDARD OPERATING PROCEDURE

SOP No. QCS-010.00 Effective date: mm/dd/yyyy
Approved by:

TITLE: **Microbiological Media Handling**

AUTHOR: ____________________

Name/Title/Department

____

Signature/Date

CHECKED BY: ____________________

Name/Title/Department

____

Signature/Date

APPROVED BY: ____________________

Name/Title/Department

____

Signature/Date

REVISIONS:

No.	Section	Pages	Initials/Date

SOP No. QCS-010.00 Effective date: mm/dd/yyyy
Approved by:

Subject: Microbiological Media Handling

10.1 PURPOSE

The purpose of this SOP is to outline a procedure to be used for handling of media in a microbiological laboratory.

10.2 PROCEDURE

Observe aseptic conditions throughout all handling steps of media in the laboratory.

1. *Holding molten media:*
   a. Maximum holding time at 48°C is 4 h.
   b. Media must be poured as soon as possible.
2. *Remelting agar:*
   a. Most media will stand remelting once in a boiling water bath.
   b. Never remelt agar more than once.
3. *Medium dispensing:*
   a. Aseptically dispense the medium into sterile tubes, flasks, or plates as desired.
   b. Immediately cover or recap the bottle to reduce the chance of contamination.
   c. Tighten the caps of the tubes in which the media have been dispensed.
4. *Pouring plates:*
   a. Media should be cooled to just less than 50°C before pouring either by using a water bath at 48°C or cooling on the bench.
   b. Immediately before pouring, the medium should be swirled gently to ensure thorough mixing.
   c. Aseptically dispense the media into sterile plates. Immediately cover or recap the bottle to reduce the chance of contamination.
   d. Plates should contain at least 5-mm depth of medium if they are to be stored before use to minimize the effect of drying that will occur during storage. In this case, the amount of medium will be more than that specified for the pour plate method.
5. *Drying plates:*
   a. Carefully place the plate under laminar air flow (LAF) with the medium-containing side up. Lift the base of the dish up and rest it on the lid.
   b. This allows excess moisture to evaporate while minimizing the possibility of contaminating the agar.
6. *Adding enrichments and supplements:*
   a. Heat-labile enrichments and supplements are added to media after sterilization to avoid heat that led to deterioration of the substance.
   b. Generally, cool sterile agar base media to approximately 45–50°C in a water bath. Aseptically add the enrichment to the medium as directed on the media container label. Swirl the flask to mix thoroughly; avoiding the formation of bubbles.
   c. Sterile broth may be cooled to room temperature before adding enrichment.
   d. Any antibiotic supplement reconstituted, but not used, should be discarded.
7. *Storage of prepared media:*
   a. Prepared media must be used on the same day of preparation.
   b. The storage period of prepared media is dependent upon the composition of the medium, the form in which it is stored and the conditions of storage.

SOP No. QCS-010.00 Effective date: mm/dd/yyyy
Approved by:

   c. Maximum 7 days, in the dark at 2–10°C, is acceptable for storage of prepared media.
   d. All media must be stored to prevent the formation of bacteriostatic and bactericidal substances (e.g., peroxides).
8. *Storage of prepared plates:*
   a. Most plates stored media side up at 2–10°C in the dark will have a maximum life of 7 days.
   b. Plates containing antibiotics have a shelf life governed by the stability of the antibiotic.
   c. Plates showing visible signs of shrinkage (drying) shall be discarded as unsuitable for use.
   d. Plates must be brought to room temperature before use to avoid any thermal shock to the bacteria.
   e. Any plates left on the bench for more than 8 h must be discarded as unsuitable for use.

10.3 REASON FOR REVISION

First time issued for your company, affiliates, and contract manufacturers.

QCS-011

YOUR COMPANY STANDARD OPERATING PROCEDURE

SOP No. QCS-011.00 Effective date: mm/dd/yyyy
Approved by:

TITLE: **General Cleaning and Sanitation of Microbiological Laboratory**

AUTHOR: ______________________________

Name/Title/Department

____

Signature/Date

CHECKED BY: ______________________________

Name/Title/Department

____

Signature/Date

APPROVED BY: ______________________________

Name/Title/Department

____

Signature/Date

REVISIONS:

No.	Section	Pages	Initials/Date

SOP No. QCS-011.00 Effective date: mm/dd/yyyy
Approved by:

Subject: General Cleaning and Sanitation of Microbiological Laboratory

11.1 PURPOSE

To establish a light level of cleaning, sanitation, and hygiene in microbiological labs.

11.2 RESPONSIBILITY

The QC cleaning staff is responsible to follow up the cleaning procedure. The QC manager/designate is responsible for the SOP compliance.

11.3 PROCEDURE

11.3.1 General Procedure

Areas to be cleaned:

- Media preparation lab
- Microbiology working lab

11.3.2 Method of Cleaning

- *Floor:* Use vacuum cleaning for dedusting. Mop with sanitation solution as described later.
- *Glass windows:* Remove dust from window corners using a vacuum cleaner; spray glint and clean with a cleaning shaft.
- *Doors:* Remove dust using a vacuum cleaner. Apply glint and clean with a synthetic soother.
- *Walls:* Remove dust using a vacuum cleaner fitted with a brush.
- *Roof and false ceiling:* Remove dust using a vacuum cleaner.
- *Table and chair*: Dedust using a dry cloth wipe with cotton soaked in any valid disinfectant according to the rotation program.
- *AC exhaust/supply/tube lights/electric switch*: Dedust using a dry cloth.
- *Cabinets*: Remove dust with a dry cloth wipe with cotton soaked in any valid disinfectant according to the rotation program.
- *Uniform*: Transfer to the laundry.
- *Trash bins*: Check and remove pharmaceutical waste from dustbins every day at the end of the shift. Verify that the dustbins are free from visual debris, disinfect the dust bin with 70% alcohol. Laboratory supervisor to verify the cleaning at the start of shift the next day.
- *Racks*: Remove dust with a dry cloth.

SOP No. QCS-011.00 Effective date: mm/dd/yyyy
Approved by:

11.3.3 Frequency

Changed every 2 weeks. Enter the data into specified log books (Attachments II and III).

	Floor	Wall	Roof ceiling	Table and Chair	Cabinet	Glass window	Uniform	Racks	Trash bins	Tube light	AC Exhaust	Electric switch
Daily	✓			✓	✓	✓	✓		✓			
Weekly		✓						✓		✓	✓	✓
Monthly		✓	✓									

Note: Make entries in specified log book (Attachments II and III).

11.3.4 List of Detergents, Disinfectants, Cleaning Items Used in Microbiology Labs

1. Tego 2000 2%
2. Ethyl alcohol 70%
3. Isopropyl alcohol 70%
4. Hydrogen peroxide 3%
5. Sodium hypochlorite 1%
6. Proseptyl 2%
7. Pera safe

11.3.5 Preparation of Tego 2000 (2% Concentration)

- Add approximately 1 L of WFI at room temperature in a clean bottle.
- Add 20 mL of Tego 2000 to the bottle.
- Complete the volume upto 1 L by sterile water at room temperature.
- Use the solution for a maximum time of 60 min after preparation for surface disinfection by swabbing or mopping or spraying.
- Make entries in specified logbook (Attachment I).

11.3.6 Preparation of 2-in-100 Solution of Proseptyl

- Measure 20 mL of the disinfectant in a 2 L measuring cylinder.
- Make up the volume of the disinfectant up to 1 L with purified water.
- Shake for 1 min.
- Make entries in a specified logbook (Attachment I).

SOP No. QCS-011.00 Effective date: mm/dd/yyyy
Approved by:

11.3.7 Preparation of Alcohol 70%

- In a 2 L volumetric flask add 1473 mL of 95% alcohol and make up the volume with sterile water.
- Shake for 1 min.
- Make entries in a specified logbook (Attachment I).

11.3.8 Preparation of 3% Hydrogen Peroxide

- Pour 100 mL of WFI in a 1 L bottle.
- Add 100 mL of 30% hydrogen peroxide.
- Make up the volume to 1 L.
- Use within 24 h.
- Make entries in a specified logbook (Attachment I).

11.3.9 Preparation of Pera Safe

- Add 16.2 g of pera safe powder in 1 L of sterile water.
- Stir until the powder dissolves in the solution and turns blue/green in color.
- Use at 20°C ± 5°C.
- Use immediately after preparation.
- Make entries in a specified logbook (Attachment I).

11.3.10 Preparation of Sodium Hypochlorite 1%

- Dilute 80 mL of sodium hypochlorite solution (12.5%) to 1-L sterile water.
- Make entries in a specified logbook (Attachment I).

11.3.11 Disinfectant Rotation Program

Change disinfectant every 2 weeks.

Disinfectant Name	Shelf Life	Use
Ethyl alcohol 70%	1 week	Hand, equipments, and utensils
Isopropyl alcohol 70%	1 week	Hand, equipments, and utensils
Pera safe	24 h	Hand, equipments, and utensils
Proseptyl 2%	24 h	Wall and floor
Hydrogen peroxide 3%	24 h	Wall and floor
Sodium hypochlorite 1%	1 week	Hand, equipments, and utensils
Tego 2000 2%	2 h	Equipment and utensils

Note: Make entries in a specified logbook (Attachment I).

11.4 REASON FOR REVISION

First time issued for the ABC Pharmaceutical Company.

11.5 ATTACHMENT FORMS

Attachment I
SOP No.:
Issued on:
Revision No.:
Initiator name: l

DISINFECTANT PREPARATION RECORD

Disinfectant name:	**Supplier name:**
Supplier lot No.:	**Lot No.:**
Manufacturing date:	**Expiry date:**
Prepared on:	**Prepared by:**

Preparation Procedure:

Weigh ____ mL or ____ g of disinfectant in ________ liter beaker

Add ____ liter of sterile water using a graduated cylinder.

Stir for ____ min. Make up the volume up to ________ liter of sterile water

Maintain the temp at ____ °C

For immediate use filter by using a sterile filtration unit,

Lot No. .... Expiry date .../..../....

Disinfectant shelf life ☐ Hour ☐ Day ☐ Week

Rotation program every ☐ Day ☐ Week ☐ Month

Disinfectant used for cleaning and disinfecting of:

☐ Micro lab office ☐ Micro lab working area

☐ Micro lab preparation room ☐ Sterility room

☐ Hand ☐ Equipments and utensils

According to Standard Operating Procedure No. [QCS0012]

Prepared by: Checked by:

Date: Date:

ABC PHARMACEUTICAL COMPANY

Attachment II
SOP No.:
Issued on:
Revision No.:
Initiator name:

DAILY REPORT
Cleaning and Sanitation Report

1. Media preparation laboratory
2. Microbiology working laboratory
3. Microbiology office

S/No.	Zones/Articles for Cleaning	Method of Cleaning	Frequency	Yes	No
1	Dustbins	Waste removal and general cleaning	Every day		
2	Floors	A. Vacuum	Every day		
		B. Sanitation with disinfectant	Every day		
3	Offices	General cleaning and sanitation	Every day		
4	Doors and windows	Cleaning	Every day		
5	Walls	Vacuum	Monthly		

Performed by:____________________ Date: ____________________

Checked by:____________________ Date: ____________________

ABC PHARMACEUTICAL COMPANY

Attachment III
SOP No.:
Issued on:
Revision No.:
Initiator name:

WEEKLY REPORT
Cleaning and Sanitation Report

1. Media preparation laboratory
2. Microbiology working laboratory
3. Microbiology laboratory office

S/No.	Zones/Articles for Cleaning	Method of Cleaning	Frequency	Yes	No
1	Table/chair	Dedust	Daily		
2	Roof/false ceiling	Vacuum	Monthly		
3	Fire extinguisher	Wet mopping	Once in a week		
4	Racks	Dry cloth	Fornightly		
5	Tube lights	General cleaning dedusting	Monthly		
6	Electrical switches	Dry cloth	Weekly		
7	A/c exhaust/ supply	Dry cloth	Weekly		

Performed by:____________________ Date: ____________________

Checked by:____________________ Date: ____________________

QCS-012

YOUR COMPANY STANDARD OPERATING PROCEDURE

SOP No. QCS-012.00 Effective date: mm/dd/yyyy
Approved by:

TITLE: **Managing Primary and Secondary Reference Standards in the Quality Control Laboratory**

AUTHOR: ______________________________

Name/Title/Department

____

Signature/Date

CHECKED BY: ______________________________

Name/Title/Department

____

Signature/Date

APPROVED BY: ______________________________

Name/Title/Department

____

Signature/Date

REVISIONS:

No.	Section	Pages	Initials/Date

SOP No. QCS-012.00 Effective date: mm/dd/yyyy
Approved by:

Subject: Managing Primary and Secondary Reference Standards in the Quality Control Laboratory

12.1 PURPOSE

To provide and define a written procedure for the management of primary and internal working standards (secondary) so as to ensure the appropriate use of standard in analytical work.

12.2 SCOPE

Includes all primary analytical reference standards, including but not limited to compendial assay, titrimetric, thermometric, and spectrophotometric standards. All materials are to be used as working (secondary or in-house) standards, including, but not limited to, raw materials and purchased materials.

12.3 RESPONSIBILITY

- It is the responsibility of the authorized person/designate to follow up the procedure and to control the reference standards. QC manager/designate will be responsible for the SOP compliance.
- It is the responsibility of each analyst to ascertain that the particular supply of primary reference standards in use is from the current lot.
- It is the responsibility of the QC manager/designate (raw material section) to review online the USP Catalogue on a bimonthly basis for any change in the status of reference standards currently in use.

12.4 PROCEDURE

1. Primary reference standards
    - Official standards (USP/BP/EP/JP) are required for all compendial monograph work. These can be purchased from US pharmacopeial convention, British pharamacopeial commission, European pharmacopeial commission, and Japanese pharmacopeia, respectively.
    - Only the current regulatory lot will be used. Current lot numbers are listed in the catalogues or cited on the web page of each pharmacopeia.
    - On receipt of a standard, the name, lot number, and date received will be entered into a log book allocated for primary reference standards only.
    - All primary standards are stored under the recommended storage conditions mentioned either on the vial of the standard or in the safety data sheet supplied with the standard. The standards are then stored in a secure location to which access is limited only to laboratory personnel having a management authority.
    - A periodical inventory (on a quarterly basis) will be carried out and documented in a log book to assure that only current lots are in the system (e.g., if current lot is "G," it means that lot "F" has expired unless the expiry date of the previous lot has not passed). Out-of-date lots of standard must be destroyed.
    - When a primary standard is needed for analytical work, it will be issued to the analyst by the relevant officer. The analyst will then sign the standard and upon return, log in the amount used and the notebook reference to actual weighings of the standard. After use, the authorized person/designate must return the standard to its secure location.

SOP No. QCS-012.00 Effective date: mm/dd/yyyy
Approved by:

2. Secondary or internal working standard
   - The equipment used is the one specified in the compendial monograph applicable to the method used to certify a potential secondary or internal working standard against a primary reference standard.
   - All the reagents to be used are those specified in the official monograph applicable to the method used to certify a potential secondary or working standard using fresh certified raw material as a sample. Perform the standard preparation and sample preparation, and determine the purity of the raw material (secondary reference vs. the primary reference standard) according to the procedure in the monograph that is being used.
   - Perform the analysis as mentioned below, using separate standard and sample weighing for each determination.
   - Calculate the assay result of the house standard versus the primary reference standard for each of the individual determinations.
   - Calculate the average and percent relative standard deviation (% RSD) for all the assays.
   - If the % RSD is 2.0 or less, then the potential secondary or house standard can be used as a reference standard for further analytical work using its average assay value as its purity.
   - Record all assay data in the notebook.
   - All the secondary or in-house standards have to be restandardized after 1 year.

12.5 PROCEDURE FOR INTERNAL WORKING STANDARDS (SECONDARY REFERENCE STANDARDS) PREPARATION

12.5.1 Sampling

After the material has been identified as a potential internal working standard, sampling of incoming raw material should be performed for characterization and certification testing. Sampling considerations should include, but are not limited to, the following:

1. The material is fresh (not older than 6 months).
2. The immediate container and packing components are intact and in good condition at the time of receipt.
3. No unfavorable, uncontrolled conditions occurred during transit, for example, exposure to excessive humidity or heat.
4. Appropriate documentation is included with the shipment, that is, dated purchase order and invoice, certificate of analysis including manufacturing date, and data for all tests required as per specifications.
5. Samples are taken under laminar flow conditions into clean, dry, sterilized containers of adequate size and provided with hermetic seal: samples are properly labeled and assigned an internal lot number.
6. The amount of sample is sufficient for all qualification tests and the quantity of material is sufficient for 1 year's usage after the qualification.

12.5.2 Testing

The samples of materials for qualification as internal working standards will be tested against current primary standards according to current standard control procedures of ABC Pharmaceutical Company and/or compendial specifications and methods.

SOP No. QCS-012.00 Effective date: mm/dd/yyyy
Approved by:

1. Define the tests and specifications that are required to evaluate the suitability of the material as an internal working standard. As a minimum, the following tests will be performed:
   a. Description
   b. Identification (by more than one specific method)
   c. Moisture
   d. Purity or potency (any other applicable purity test impurities, degradation products)
   e. Any other applicable stability indication tests
2. The potency of each sample lot will be tested on "as is basis" in triplicate, on two separate days, and preferably by the same analyst, which means a total of six tests per sample is required.
3. Evaluate statistically the results for potency/purity as per ABC Pharmaceutical Company's standard control procedures. If the individual results are valid, calculate the average value and assign the average potency/purity to the material.

   Note: The potency/purity value obtained must be statistically comparable to data shown in the supplier's certificate of analysis.

a. Review the data to confirm that the material is suitable, assign the potency/purity value determined by the assays, and designate the material as an internal working standard.
b. Issue a signed and dated certificate of analysis for this internal working standard.
c. Keep all documentation, data safety precautions, and test results in a file designated as part of the internal working standard program for review or use by an authorized personnel.
d. Assign a 1-year expiry date to internal working standards. At the end of the expiry period, any quantity remaining of the lot must be discarded. On an exceptional basis, some secondary reference standards may be recertified for an additional 1 year period. The standards must not be used beyond 2 years from the original date of certification.
e. The reason for assigning 1 year expiry date for the secondary reference standards is the fact that the internal working standard maximum of 2 g/vial is standardized at a time and the frequency of usage ensures that the standard does not exist beyond 1 year. Besides this, the history of the material also shows from manufacturing suppliers certificate that they are stable for a period of up to 3–4 years which also supports the validity of these secondary standards for 1 year under controlled temperature.
f. Review the data to confirm that the material is suitable, assign the potency/purity value determined by the assays, and designate the material as an internal working standard (secondary reference standard).

12.5.3 Subdivision and Labeling

When the certification of the internal working standard is completed, it should be subdivided into small quantities, not exceeding 2 g/vial (for moisture-sensitive material, it should not exceed 200 mg/vial). The container should be of amber glass or plastic (if compatible with the material), clean and dry, with dry, hermetic closures, and the subdivision should be performed by trained personnel under controlled conditions. Each vial should be labeled to show:

SOP No. QCS-012.00 Effective date: mm/dd/yyyy
Approved by:

Internal working standard:

1. Material name
2. Assay (as is basis)
3. Lot numbers
4. Expiry date
5. Storage condition, if any

12.5.4 Expiry Date

The expiry date of the internal working standards on the vial/s label is mentioned as MM/YYYY, which means that the standard can be used till the end of the month of the corresponding year. For example, if the expiry date is given 07/2013, this would mean that the standard can be used till July 31, 2013 and will be considered expired on August 01, 2013.

The authorized person/designate will conduct an inspection at the end of each month to check any expired standard. The expired standard will be removed immediately and documented in the log book.

12.5.5 Storage Conditions

Unless otherwise indicated, internal working standards must be stored in a cool and dry place away from heat and protected from light. Some reference standards may be required to be placed in individual plastic bags inside a desiccator placed in a refrigerator.

12.5.6 Usage

1. Internal working standards will be used for the testing of routine production batches whenever the methodology calls for the use of a "reference standard."
2. Calculations will take into consideration the labeled potency/purity of the internal standard.
3. In some special cases, for example, to resolve analytical discrepancies, a primary reference standard can also be used. Internal working standards cannot be used to standardize other internal working standards. Only primary standards will be used for this purpose.

Note:

- The handling of primary and secondary reference standards will be maintained through a register. The standards for new raw materials will be added in the register on an ongoing basis.
- Only sufficient quantity of primary standards will be purchased and long-term storage should be avoided.
- Secondary reference standards which are stored under nitrogen will be used only once and the remaining quantity will be discarded.
- Quantity in each vial for other secondary reference standards should be sufficient for being used three times and it should be discarded afterward.
- Primary reference standards, including controlled drug substances, will be destroyed according to the SOP for destruction (disposal) of samples after complete testing.

SOP No. QCS-012.00 Effective date: mm/dd/yyyy
Approved by:

12.6 REASONS FOR REVISION

Date		Raw Material Test Method		
No.	**Documentation Description**	**Code No.**	**Issue Date**	**Supersedes Date**
1	Acetazolamide			
2	Ascorbic acid			
3	Atropine sulfate			
4	Ascorbic acid 90% granulated			
5	Acyclovir			
6	Amoxicillin trihydrate fine powder			
7	Ampicillin (anhydrous)			
8	3-Amino-2-(3,4,5 trimethoxybenzyl)acrylomitride			
9	Atenolol			
10	Beta-sitosterol (synthetine)			
11	Beclomethasone dipropronate			
12	Betamethosone valerate			
13	Bisacodyl micronized			
14	4-Bron-2,2 siphenyl butyronitride			
15	2-(4-Butylphenyl)-propronic acid			
16	Calcium panthothenate			
17	Cephalexin			
18	Carbinoxamine maleate pellets			
19	Cefaclor, compacted			
20	Caffeine			
21	Cefaclor, fine			
22	Cephradine			
23	Cephalexin monohydrate (compacted)			
24	Cephalexin monohydrate (fine powder)			
25	Clidimiun bromide			
26	Chlordiazepoxide			
27	Chlortetracycline hydrochloride			
28	Chlorpheniramine maleate			
29	7-Chloro-1,3-dihydro-5-phenyl-21 + 1, 4 benzodiazepin-2-one 4 oxide			
30	Ciprofloxacin HCl			
31	Cisapride			
32	Ciprofloxacin impurity I			
33	Cimetidine			
34	Cloxacillin sodium			
35	Clibetasol profinate			
36	Clibetasol profinate related compound A			
37	Cortisone acetate			
38	4-[2-(5-Chloro-2 methoxybenzamide)ethyl]benzine sulfonamide			
39	Cyclohexa-1,4-dienylglycine			
40	Cyanocobalamin			
41	Dibucaine HCl			

SOP No. QCS-012.00 Effective date: mm/dd/yyyy
Approved by:

(continued)

Date		Raw Material Test Method		
No.	**Documentation Description**	**Code No.**	**Issue Date**	**Supersedes Date**
42	Diclofenac related compound A			
43	Dextromethorphan HBr			
44	Dextrose, anhydrous			
45	Diphenhydramine HCl			
46	Doxycycline hyclate			
47	Efrianhydrotetracycline			
48	Ephedrine sulfate			
49	Ergosterol			
50	Erythromycin ethyl succinate USP 23			
51	Erythromycin stearate			
52	Erythromycin (base)			
53	Ergocalceferol			
54	Ethyl-2-(2-methylbenzhydryloxy) ethyl dimethyl ammonium chloride			
55	Famotidine			
56	Famotidine CRS			
57	Famotidine impurity A CRS			
58	Famotidine impurity B CRS			
59	Folic acid			
60	Furosemide			
61	Fluoroquindonic acid			
62	Glibenclamide micronized powder			
63	Gramicidin			
64	Gentamicin sulfate			
65	Griseofulvin			
66	Guaracol			
67	Guaraphenesin			
68	Haloperidol			
69	Hydrocortisone acetate			
70	Hydrocortisone			
71	Hyoscine hydrobromide			
72	1-(4-Hydroxy-3-methylphenyl)-2-(*t*-butylamine)ethanol			
73	3-Hydroxy-1-methylquinuclidinium bromide			
74	Ibuprofen			
75	4-Isobutylacetophenone			
76	Indomethocin			
77	Indigotindisulfonate sodium			
78	Lidocaine HCl			
79	Mebendazole			
80	Miconazole nitrate			
81	Metoclopramide hydrochloride			
82	Metronidazole			
83	Metronidazole benzoate			
84	2-(2-Methylbenzhydryloxy)ethyl-*N*-methylammonium HCl			

continued

SOP No. QCS-012.00 Effective date: mm/dd/yyyy
Approved by:

(continued)

Date		Raw Material Test Method		
No.	**Documentation Description**	**Code No.**	**Issue Date**	**Supersedes Date**
85	Methyl-*N*-4[-(5-chloro-2-methoxy benzamido)ethyl] benzene sulfonyl carbonate			
86	Neomycin sulfate			
87	Nifedipine nitrophenylphyridine analog			
88	Niacinamide			
89	Norfloxacin			
90	Nystatin			
91	Omeprazole			
92	Orphenadrine citrate			
93	Oxymetazoline HCl			
94	Oxytetracycline HCl			
95	Paracetamol			
96	Phenylpropanolamine HCl pellets			
97	Prednisone			
98	Prednisolone			
99	Promethazine HCl			
100	Propranolol HCl			
101	Pseudoephedrine HCl			
102	Pyridoxine HCl			
103	3-Quinuclidingl benzilate			
104	Ranitidine hydrochloride			
105	Ranitidine-related compound A			
106	Ranitidine-related compound B			
107	Ranitidine-related compound C			
108	Riboflavin			
109	Salicylamide			
110	Salbutamol sulfate			
111	Sennosides			
112	Simethicone			
113	Sulfamilamide			
114	Sulfamethoxazole-N4-glucoside			
115	Terfenadine			
116	Tetracycline HCl			
117	Terfenadine-related compound A			
118	Thiamine HCl			
119	Trimethoprin			
120	Triprolidine HCl			
121	Triprolidine HCl Z-isomer			
122	Xylometazoline hydrochloride			

SOP No. QCS-012.00 Effective date: mm/dd/yyyy
Approved by:

12.7 ATTACHMENT FORM

Attachment No. I
SOP No.: QCS-
Issued on:
Revision No.:
Initiator signature:

S. No.	Primary Standard	Reference Method	Catalogue Number
1	3-Amino-2-(3,4,5 trimethoxybenzyl)acrylomitride		
2	Acetazolamide		
3	Acyclovir		
4	Amoxicillin trihydrate fine powder		
5	Ampicillin (anhydrous)		
6	Ascorbic acid		
7	Ascorbic acid 90% granulated		
8	Atenolol		
9	Atropine sulfate		
10	2-(4-Butylphenyl)-propionic acid		
11	4-Bron-2,2 siphenyl butyronitride		
12	Beclomethasone dipropionate		
13	Betamethosone valerate		
14	Bisacodyl micronized		
15	B-Setosterol (synthetine)		
16	4-[2-(5-Chloro-2 methoxybenzamide)ethyl]benzine sulfonamide		
17	7-Chloro-1,3-dihydro 5 phenyl-21 + 1, 4 benzodiazepin-2-one 4 oxide		
18	Caffeine		
19	Calcium panthothenate		
20	Carbinoxamine maleate pellets		
21	Cefaclor, compacted		
22	Cefaclor, fine		
23	Cephalexin		
24	Cephalexin monohydrate (compacted)		
25	Cephalexin monohydrate (fine powder)		
26	Cephradine		
27	Chlordiazepoxide		
28	Chlorpheniramine maleate		
29	Chlortetracycline hydrochloride		
30	Cimetidine		
31	Ciprofloxacin HCl		
32	Ciprofloxacin impurity I		
33	Cisapride		
34	Clobetasol propionate		
35	Clobetasol propionate-related compound A		
36	Clidinium bromide		

continued

SOP No. QCS-012.00 Effective date: mm/dd/yyyy
Approved by:

(continued)

S. No.	Primary Standard	Reference Method	Catalogue Number
37	Cloxacillin sodium		
38	Cortisone acetate		
39	Cyanocobalamin		
40	Cyclohexa-1,4-dienylglycine		
41	Dextromethorphan HBr		
42	Dextrose, anhydrous		
43	Dibucaine HCl		
44	Diclofenac-related compound A		
45	Diphenhydramine HCl		
46	Doxycycline hyclate		
47	Efrianhydrotetracycline		
48	Ephedrine sulfate		
49	Ergocalceferol		
50	Ergosterol		
51	Erythromycin (base)		
52	Erythromycin ethyl succinate USP 23		
53	Erythromycin stearate		
54	Ethy-2-(2-methylbenzhydroxyloxy) ethydimethyl ammonium chloride		
55	Famotidine		
56	Famotidine CRS		
57	Famotidine impurity A CRS		
58	Famotidine impurity B CRS		
59	Fluoroquindonic acid		
60	Folic acid		
61	Furosemide		
62	Gentamicin sulfate		
63	Glibenclamide micronized powder		
64	Gramicidin		
65	Griseofulvin		
66	Guaracol		
67	Guaifenesin		
68	1-(4-Hydroxy-3-methylphenyl)-2-(*t*-butylamine)ethanol		
69	3-Hydroxy-1-methylquinuclidinium bromide		
70	Haloperidol		
71	Hydrocortisone		
72	Hydrocortisone acetate		
73	Hyoscine hydrobromide		
74	4-Isobutylacetophenone		
75	Ibuprofen		
76	Indigotindisulfonate sodium		
77	Indomethacin		
78	Lidocaine HCl		
79	2-(2-Methylbenzhydryloxy)ethyl-*N*-methylammonium HCl		
80	Mebendazole		

SOP No. QCS-012.00 Effective date: mm/dd/yyyy
Approved by:

(continued)

S. No.	Primary Standard	Reference Method	Catalogue Number
81	Methyl-*N*-4[-(5-chloro-2-methoxy benzamido)ethyl] benzene sulfonyl carbonate		
82	Metoclopramide hydrochloride		
83	Metronidazole		
84	Metronidazole benzoate		
85	Miconazole nitrate		
86	Neomycin sulfate		
87	Niacinamide		
88	Nifedipine nitrophenylphyridine analog		
89	Norfloxacin		
90	Nystatin		
91	Omeprazole		
92	Orphenadrine citrate		
93	Oxymetazoline HCl		
94	Oxytetracycline HCl		
95	Paracetamol		
96	Phenylpropanolamine HCl pellets		
97	Prednisolone		
98	Prednisone		
99	Promethazine HCl		
100	Propranolol HCl		
101	Pseudoephedrine HCl		
102	Pyridoxine HCl		
103	3-Quinuclidingl benzilate		
104	Ranitidine hydrochloride		
105	Ranitidine-related compound A		
106	Ranitidine-related compound B		
107	Ranitidine-related compound C		
108	Riboflavine		
109	Salbutamol sulfate		
110	Salicylamide		
111	Sennosides		
112	Simethicone		
113	Sulfamethoxazole-N4-glucoside		
114	Sulfamilamide		
115	Terfenadine		
116	Terfenadine-related compound A		
117	Tetracycline HCl		
118	Thiamine HCl		
119	Trimethoprim		
120	Triprolidine HCl		
121	Triprolidine HCl Z-isomer		
122	Xylometazoline hydrochloride		

QCS-013

YOUR COMPANY STANDARD OPERATING PROCEDURE

SOP No. QCS-013.00 Effective date: mm/dd/yyyy
Approved by:

TITLE: **Handling of Raw/Packaging Material and Finished Product Samples in the Reference Room**

AUTHOR: ____________________

Name/Title/Department

____

Signature/Date

CHECKED BY: ____________________

Name/Title/Department

____

Signature/Date

APPROVED BY: ____________________

Name/Title/Department

____

Signature/Date

REVISIONS:

No.	Section	Pages	Initials/Date

SOP No. QCS 013.00 Effective date: mm/dd/yyyy
Approved by:

13.1 PURPOSE

To provide a stepwise procedure for laboratory analysts to arrange, retain, and assign storage duration to the packaging materials, raw materials, and finished product reference samples and monitor the environmental conditions of the reference sample room.

13.2 RESPONSIBILITY

1. Manager QC/designate will be responsible for handling the reference samples for raw material.
2. It is the responsibility of the physical testing lab officer/designate to follow the described procedure for maintaining and keeping reference samples of finished products.
3. QC packaging officer/designate will be responsible for maintaining the packaging material reference samples.
4. The physical testing lab officer will be further responsible for monitoring the environmental conditions of the storage areas for reference samples of raw/packaging materials and finished products.
5. Manager/director quality control will be responsible for the SOP compliance.

13.3 SCOPE

This SOP is applicable to reference samples of all raw and packaging materials and finished products manufactured and packaged at ABC Pharmaceutical Company as per FDA GMP guidelines, EU-GMP guideline (Annex. 19), and the requirements of local drug rules.

13.4 PROCEDURE

1. Arrange samples of raw and packaging materials and finished products appropriately according to the batch number.
2. Keep all samples/products well segregated.
3. Monitor the temperature and the humidity of the room through temperature/humidity monitoring chart on a daily basis. In case of a deviation, inform the QC manager to call the maintenance staff for corrective action and implementation.
4. Maintain separate files for temperature and humidity records.
5. After checking the temperature and the humidity, sign the charts with date.
6. Reference samples from each batch of finished products will be retained for at least 1 year after the expiry date. The raw material reference samples will be retained for 3 years after the shelf life of the finished product.
7. After the retention period is over, the samples will be destroyed as per site SOP ABC-000.

Note: Room temperature range: 15–25°C; Humidity range: 35–65%.

13.5 REASON FOR REVISION

First time issued for your company, affiliates, and contract manufacturers.

QCS-014

YOUR COMPANY STANDARD OPERATING PROCEDURE

SOP No. QCS-014.00 Effective date: mm/dd/yyyy
Approved by:

TITLE: **Stability Testing Program**

AUTHOR: ______________________________

Name/Title/Department

____

Signature/Date

CHECKED BY: ______________________________

Name/Title/Department

____

Signature/Date

APPROVED BY: ______________________________

Name/Title/Department

____

Signature/Date

REVISIONS:

No.	Section	Pages	Initials/Date

SOP No. QCS-014.00 Effective date: mm/dd/yyyy
Approved by:

14.1 PURPOSE

To ensure that marketed products and samples, at the customer's end, meet label claim, regulatory requirements, and ABC Pharmaceutical Company's standards during stated expiration dating.

14.2 RESPONSIBILITY

It is the responsibility of the stability supervisor to follow up the procedure; the R&D manager is responsible for the SOP compliance.

14.3 SCOPE

1. Stability program provides analytical data for active ingredients and other measurable parameters that are affected by time and relevant conditions.
2. Stability test will be performed
   a. On routine production batches
   b. If relevant changes have been made in the manufacturing procedure
   c. In the event of significant changes in primary packaging
   d. Where new grade excipients which might affect stability are to be employed
   e. In problem cases
   f. On the first three representative trial batches of newly developed formulation
3. This program also provides data which can support the revision of expiration date, manufacturing excess, container change, and so on.

14.4 PROCEDURE

14.4.1 Selection of Batches

1. Stability information for accelerated and long-term testing is to be provided on three batches of the same formulation and dosage form in the containers and closure proposed for marketing.
2. The long-term testing should cover at least 12 months duration at the time of submission.
3. Data on associated formulations or packaging may be submitted as supportive information. The first commercial batch manufactured post approval, if not submitted in the original registration application, should be placed on accelerated and long-term stability studies using the same stability protocols as in the approved drug application.

14.4.2 Test Procedures and Test Criteria

The testing should cover those features that are susceptible to change during storage and are likely to influence quality, safety, and/or efficacy.

14.4.3 Specifications

1. Limits of acceptance should relate to the release limits (where applicable).
2. The shelf life specification could allow acceptable and justifiable derivations from the release specification based on the stability evaluation and the changes observed on storage.
3. It will need to include specific upper limits for degradation products, the justification for which should be influenced by the levels observed in batches used in preclinical studies and clinical trials (where applicable).

SOP No. QCS-014.00 Effective date: mm/dd/yyyy
Approved by:

14.4.4 Storage Test Conditions/Frequency

The length of the studies and the storage conditions should be sufficient to cover storage, shipment, and subsequent use (e.g., reconstitution or dilution as recommended in the labeling). See Table 14.1 for accelerated and long-term storage conditions and minimum times. An assurance that long-term testing will continue to cover the expected shelf life should be provided.

TABLE 14.1
Accelerated and Long-Term Storage Conditions and Minimum Times

For Solid Sample

Type of Stability Study	B. No.	Temperature and Relative Humidity	Months Till Complete Shelf Lives				
	(a)	30°C ± 2°C	1st year	2nd year	3rd year	4th year	5th year
Long term	(b)	(60% RH)					
	(c)						
			0 3 6 9 12	18 24	36	48	60
Accelerated or short term (I)	(d)	40°C ± 2°C	Two				
	(e)	(75 ± 5% RH)		3	6		
	(f)						
(II)	(g)	50°C ± 2°C	One				
	(h)			3M			
	(i)						
			Two				
For products to be kept refrigerated for long term	(j)	20°C ± 2°C		3	6		
	(k)	25°C ± 2°C					
	(l)						

For Liquid Sample

Type of Stability Study	B. No.	Temperature and Relative Humidity	Months Till Complete Shelf Lives				
	(a)	30°C ± 2°C with	1st year	2nd year	3rd year	4th year	5th year
Long term	(b)						
	(c)						
			0 3 6 9 12	18 24	36	48	60
Accelerated or short term (I)	(d)	40°C ± 2°C with	Two				
	(e)			3	6		
	(f)						
(II)	(g)	50°C ± 2°C	One				
	(h)			3			
	(i)						
			Two				
For products to be kept refrigerated for long term	(j)	20°C ± 2°C		3	6		
	(k)	25°C ± 2°C					
	(l)						

SOP No. QCS-014.00 Effective date: mm/dd/yyyy
Approved by:

14.4.4.1 Long-Term Testing

Long-term tests must be carried out on representative batches using the type of packages intended for marketing. A 12-month long-term stability data should be provided at the time of submission. In this case, the storage is carried out at 30°C ± 2°C/60% RH for solid phase and 30°C ± 2°C for liquid phase. This should be done for all types of drug products and the study should cover not less than 12 months duration at the time of submission for registration, provided that the manufacturer has to give a written assurance that long-term testing will continue to cover the expected shelf life completely.

In this type of study, the observations and testing are done first at zero time, every 3 months during the first year, every 6 months during the second year, and then every year till the end of the shelf life.

14.4.4.2 Short-Term or Accelerated Stability Testing

a. A 6-month accelerated stability data should be provided at the time of submission for registration which should be carried out at the following storage conditions: 40°C ± 2°C/75 + 5% RH for solid sample and 40°C ± 2°C for liquid sample and 3 months date for samples stored at 50°C ± 2°C.
b. For products to be stored for long term under refrigerated condition, the accelerated testing should be conducted at

25°C ± 2°C

20°C ± 2°C

c. In these accelerated stability studies, the storage under high humidity applies particularly to solid dosage forms which are not packed in containers designed to provide a permanent barrier to water loss. For such formulations, the relative humidity should be lowered by 10–20%.
d. For some dosage forms, especially liquid and semisolid dosage forms, the study design may also need to consider low-temperature freezer (–10°C to –20°C) and refrigerator (2°C to 8°C), but usually, accelerated studies are less suitable for semisolids and heterogeneous formulations, for example, emulsions.
e. Other storage conditions in stability studies are allowable if justified. The heat-sensitive drug products should be stored under an alternative lower-temperature condition, if justified by the manufacturer, which will eventually become the designated long-term storage temperature. Studies at low temperature that affects and changes the finished product's properties should be avoided.
f. For products such as solutions, suspensions, and so on contained in packs designed to provide a permanent barrier to water loss, specific storage, under condition of high relative humidity, is not necessary.

14.4.4.3 For (Routine) Ongoing Stability Studies

14.4.4.3.1 On Retained Batches Obtained from Current Production

a. Zero-time result will be taken from the QC department "release result."
b. Every year, for some products, one batch will be selected from the sample reference room in QC, which is near to expiry for analysis by the stability staff.

14.4.4.3.2 Stability Data for Supplemental Changes

Stability data submitted in support of supplemental changes for an existing drug product may be generated with samples at the accelerated testing conditions and long-term testing conditions.

SOP No. QCS-014.00 Effective date: mm/dd/yyyy
Approved by:

14.4.4.4 Product Formula and/or Process Change

Stability samples are required from the first product lot following a formula or manufacturing process change which may have stability significance.

For example, a change in active ingredient excess would be a *formula* change having stability significance.

14.4.4.5 Packaging Materials

The testing should be carried out in the final packaging proposed for marketing. Additional testing of unprotected drug product can form a useful part of the stress testing and pack evaluation, as can studies carried out in other related packaging materials in supporting the definitive pack(s).

14.4.5 Evaluation

A systematic approach should be adopted in the presentation evaluation of the stability information which should cover, as necessary, physical, chemical, biological, and microbiological quality characteristics, including particular properties of the dosage form (e.g., dissolution rate for oral solid dose forms).

An acceptable approach for quantitative characteristics that are expected to decrease with time is to determine the time at which the 95% one-sided confidence limit for the mean degradation curve intersects the acceptable lower specification limit. If analysis shows that the batch-to-batch variability is small, it is advantageous to combine the data into one overall estimate and this can be done by first applying appropriate statistical tests (e.g., p values for level of significance of rejection of more than 0.25) to the slopes of the regression lines and zero time.

14.5 SIGNIFICANT CHANGES AT THE ACCELERATED CONDITIONS

Significant changes at the accelerated conditions are defined as:

a. A 5% potency loss from initial assay value of a batch. All obtained values for all batches should be combined and treated by appropriate statistical tests at 95% confidence limit.
b. Any specific degradation exceeding its specification limit.
c. Exceeding in its pH limits.
d. Dissolution exceeding the specification limits for 12 caps or tablets.
e. Failure to meet specifications for appearance and physical properties such as change in color, phase separation, resuspend ability, delivery per actuation, caking, hardness, and so on.
f. Microbiological specification limit.
g. The standard conditions for light affects certain preparations. It is important to observe effects caused by their exposure to light.

14.6 STABILITY PROTOCOLS/EXECUTION

After the product is selected, the stability officer will raise the protocol form and enter the following:

1. Product name
2. Batch number
3. Shelf life
4. Reason for study

SOP No. QCS-014.00 Effective date: mm/dd/yyyy
Approved by:

5. Details of packaging commodities
6. Date of commencement
7. Date of completion
8. Stations
9. Intervals
   a. After the protocol is finalized, samples of 30, 40/50°C are arranged in a basket and kept in respective ovens and mentioned in the computer list. For any deviation or variation from the protocols, permission should be taken from the director quality affairs.
   b. Raise the stability card and enter the following:
      i. Product name
      ii. Batch number
      iii. Initial results with manufacturing excess
      iv. Testing intervals and stations
      v. Date of commencement
      vi. Date of completion
      vii. Details of packaging commodities
   c. By the last day of each month, the computer list for products falling for testing next month is provided by the stability lab officer for daily planning purpose.
   d. On the due date, the analysis is done and the results are entered on the stability card.
   e. The analyst will also mention STM No., TIME, and REF.REC.
   f. Calculation for rate of degradation:

$$\%\,\text{Degradation/month} = \frac{\text{initial result} - \text{result (specified)}}{\text{months}}$$

   For example, at 3 months

$$= \frac{102 - 98}{3} = 1.33\% \,/\, \text{month}$$

   g. If any abnormal behavior is observed, it should be discussed and brought to the knowledge of the R&D manager and the director quality affairs without any delay.
   h. Stability cards will be reviewed by the designated stability officer/PDL manager every month.
   i. Any alteration in the test interval will be done if necessary by the stability officer with consultation of the director quality affairs.
   j. Any abnormality observed by the stability officer that could affect the shelf life should be considered *important* and should be communicated immediately to the R&D manager and the director quality assurance.
   k. If the stability data fall outside the specifications, established for the previously approved expiration dating period, the applicant should perform an investigation to determine the probable cause of the failure if it is determined that the storage condition is the cause for stability failure; the applicant may shorten the expiration dating. Other measures (e.g., more protective container/closure or product reformulation) may be considered through prior approval supplement.

SOP No. QCS-014.00 Effective date: mm/dd/yyyy
Approved by:

l. Reports of the product which has potentials to extend or reduce the shelf life or requires any adjustment in excesses are to be forwarded to the director quality assurance quarterly. It consists of product name, batch number, date of commencement, date of completion, and reason for study.

Note: After the completion of the stability testing period, the remaining samples of that particular batch should be removed to avoid unnecessary congestion at the respective stations.

14.6.1 Color

A product on stability can undergo two types of color changes:

1. If the product is changed from the original color to another color, for example, from red to yellow or from blue to purple, the analyst should describe the color change as mentioned in the above example.
2. If the product maintains the original color but with lighter or darker shade of that color, there could be three types of changes, that is

   Insignificant: A change which is not readily noticeable
   Significant: A change which is readily noticeable
   Severe: A gross change

   The lighter or darker shade can be described as follows:

   Fading of color: This term will be used to indicate the change from the original to a lighter shade of the color.
   Darkening of color: This term will be used to indicate the change from the original to a darker shade of the color

14.6.2 Odor

The following phrases will be used to describe the change of odor.

Characteristic odor: Represents the original odor of the product
Faint characteristic odor: If the odor is less intense than usual
Off odor: If the odor is different from the characteristic original odor

Off odor can be subdivided into three types of description depending upon the degree of intensity: insignificant, significant, and severe to indicate not readily noticeable, readily noticeable, and gross change, respectively.

14.6.3 Taste

The following phrases will be used to describe the changes in taste:

Characteristic taste: Represents the original taste of the product
Off taste: If the taste has changed from the original characteristic taste to other taste

The term off taste will be used with the adjectives such as insignificant, significant, and severe to indicate not readily noticeable, readily noticeable, or gross change, respectively.

SOP No. QCS-014.00 Effective date: mm/dd/yyyy
Approved by:

14.6.4 Clarity

Clarity can be described as follows:

Clear	If the liquid contains no insoluble matter.
Suspended particle	If there is a separation of some insoluble material but not in the form of precipitates.
Precipitates	If there is a substantial separation of insoluble matter. There can be some sedimentation also.
Cloudy precipitate	If the precipitates are very fine and give the appearance of cloud. In this case, there can be some sedimentation.
Heavy precipitate of fine particles	Generally no sedimentation is observed.
Sedimentation	When the supernatant liquid is clear but there is some sediment at the bottom.

The terms mentioned above will be used with the adjectives such as insignificant, significant, and severe to indicate not readily noticeable, readily noticeable, and gross change, respectively.

1. For the required tests of any product, the stability staff should follow the specification at the end of the shelf life based on the finished product specification.
2. For registration, follow documentation as under:
   a. The protocol of both long-term and accelerated stability studies for any drug products should be in the original application files.
      Analytical test procedures should be stability-indicating (i.e., detects active as well as level of degradation components) and should be fully validated.
   b. The incomplete results from long-term stability can be submitted along with the completed short-term (accelerated) results; then the remaining results for long-term stability will be accepted after completion and will be added to the product original file after the approval of abbreviated new drug application (ANDA).
   c. Specification at the end of the shelf life should be included.
   d. The first page: authorization person for signature, brand name, and strength.
   e. The second page: for protocol *index* only, batch number, storage conditions, and packing material (PM).
   f. The third page and inverse should be arranged to include the following:
      - Experimental, detailed
      - Results with tables and graphs

14.7 REASON FOR REVISION

First time issued for your company, affiliates, and contract manufacturers.

QCS-015

YOUR COMPANY STANDARD OPERATING PROCEDURE

SOP No. QCS-015.00 Effective date: mm/dd/yyyy
Approved by:

TITLE: **Control of Controlled Substances (Raw Material, Bulk and Finished Products)**

AUTHOR: ______________________________

Name/Title/Department

____

Signature/Date

CHECKED BY: ______________________________

Name/Title/Department

____

Signature/Date

APPROVED BY: ______________________________

Name/Title/Department

____

Signature/Date

REVISIONS:

No.	Section	Pages	Initials/Date

SOP No. QCS-015.00 Effective date: mm/dd/yyyy
Approved by:

15.1 PURPOSE

The purpose of this SOP is to describe the procedure for the handling of controlled substances.

15.2 RESPONSIBILITY

The in-process staff of the QA department and the QC laboratory analysts are responsible to follow up the procedure. The QA and QC managers/supervisors are responsible for the SOP compliance.

15.3 PROCEDURE

15.3.1 List of Controlled Substances at the ABC Pharmaceutical Company

Diazepam
Phenobarbitone
Di-phenoxyate HCl
Codeine phosphate

15.3.2 Raw Materials

i. The raw materials are sampled by the QA sampler as per quantity specified in the raw material sampling procedure (sufficient for testing only). A blue requisition form approved by the QA director will be forwarded to the store.
ii. The sample is marked as "CDS" material and handed over to the QA officer in the stores.
iii. The QA officer will then hand over the sample to the QC supervisor accompanied with a receiving report.
iv. The QC supervisor (material receiving) makes sure that the materials are entered in the materials receiving register. The samples are kept under the custody of the QC manager.
v. The QC supervisor (raw material) will hand over the material to the analyst for testing.
vi. The samples will be kept under safe custody until the analyses is completed.
vii. After completion of the test, the leftover quantity (if any) is handed over to the QC manager.
viii. The leftover quantity is destroyed by burning in a furnace at 800°C.
ix. The QC manager will then verify the destruction of the material.

15.3.3 Bulk Products

i. The bulk products are sampled by the QA inspector as per sampling quantity specified in the bulk products sampling procedure (sufficient for testing only).
ii. The QA inspector will hand over the sample to the QC supervisor (material receiving).
iii. The QC supervisor (material receiving) makes sure that the materials are entered in the materials receiving register. The samples will be kept under the the custody of the QC manager.
iv. The QC supervisor (bulk material) will hand over the materials to the analyst for testing.
v. The sample will again be kept under safe custody until the analyses is completed.
vi. After completion of the test, the leftover quantity, if any, is handed over to the QC manager.
vii. The leftover sample is destroyed. Solid samples are soaked in water and drained. The liquid samples are directly drained.

SOP No. QCS-015.00 Effective date: mm/dd/yyyy
Approved by:

15.3.4 Finished Products

i. Finished products are sampled by the QA inspector as per the sampling quantity specified in the finished products sampling procedure (sufficient for testing only).
ii. The QA inspector will hand over the sample to the QC supervisor (material receiving).
iii. The QC supervisor (material receiving) makes sure that the materials are entered in the materials receiving register. The samples will be kept under the custody of the QC manager.
iv. The QC supervisor (finished products) will then hand over the materials for testing.
v. The sample will be under safe custody until the analysis is completed.
vi. After completion of the test, the leftover quantity is preserved in a file sample room for the period of shelf-life plus one year.

15.4 REASON FOR REVISION

First time issued for your company, affiliates, and contract manufacturers.

QCS-016

YOUR COMPANY STANDARD OPERATING PROCEDURE

SOP No. QCS-016.00 Effective date: mm/dd/yyyy
Approved by:

TITLE: **Rounding of Significant Numbers**

AUTHOR: ______________________________

Name/Title/Department

____

Signature/Date

CHECKED BY: ______________________________

Name/Title/Department

____

Signature/Date

APPROVED BY: ______________________________

Name/Title/Department

____

Signature/Date

REVISIONS:

No.	Section	Pages	Initials/Date

SOP No. QCS-016.00 Effective date: mm/dd/yyyy
Approved by:

16.1 PURPOSE

To provide a procedure for the rounding of significant numbers.

16.2 RESPONSIBILITY

It is the responsibility of all analysts, microbiologists, QA in-process, validation, maintenance (calibration staff), and production and packaging staff to follow the procedure. All corresponding department's managers are responsible for SOP compliance. The QA director will be responsible for overall systems compliance.

16.3 PROCEDURE

16.3.1 USP Criteria

Where limits are expressed numerically herein, the upper and lower limits of a range include the two values themselves and all intermediate values, but no values outside the limits. The limits expressed in monographic definitions and tests, regardless of whether the values are expressed as percentages or as absolute numbers, are considered significant to the last digit shown.

16.3.2 Equivalence Statements in Titrimetric Procedures

The directions for titrimetric procedure conclude with a statement of the weight of the analyte that is equivalent to each milliliter of the standardized titrant. In such an equivalence statement, it is to be understood that the number of significant figures in the concentration of the titrant corresponds to the number of significant figures in the weight of the analyte. Blank corrections are to be made for all titrimetric assays where appropriate.

16.3.3 Tolerances

The limits specified in the monographs for pharmacopeial articles are established with a view to the use of these articles as drugs, except where indicated otherwise. The use of the molecular formula for the active ingredient(s) named in defining the required strength of a pharmacopeial article is intended to designate the chemical entity of entities, as given in the complete chemical name of the article, having absolute (100) purity.

A dosage form shall be formulated with the intent to provide 100% of the quantity of each ingredient declared on the label. Where the content of an ingredient is known to decrease with time, an amount in excess of that declared on the label may be introduced into the dosage form at the time of manufacture to assure compliance with the content requirements of the monograph throughout the expiration period. The tolerances and limits stated in the definitions in the monographs for pharmacopeial articles allow for such overages and for analytical error, for unavoidable variations in manufacturing and compounding, and for deterioration to an extent considered acceptable under practical conditions.

The specified tolerances are based upon such attributes of quality as might be expected to characterize an article produced from suitable raw materials under recognized principles of good manufacturing practice.

The existence of compendial limits or tolerances does not constitute a basis for a claim that an official substance that more nearly approaches 100% purity "exceeds" the pharmacopeial quality.

SOP No. QCS-016.00 Effective date: mm/dd/yyyy
Approved by:

Similarly, the fact that an article has been prepared to closer tolerances than those specified in the monograph does not constitute a basis for a claim that the article "exceeds" the pharmacopeial requirements.

16.3.3.1 Interpretation of Requirements

Analytical results observed in the laboratory (or calculated from experimental measurements) are compared with stated limits to determine whether there is conformance with compendial assay or test requirements. The observed or calculated values usually will contain more significant figures than there are in the stated limit, and an observed or calculated result is to be rounded off to the number of places that are in agreement with the limit expression by the following procedure (*Note:* Limits, which are fixed numbers, are not rounded off).

When rounding off is required, consider only one digit in the decimal place to the right of the last place in the limit expression. If this digit is <5, it is eliminated and the preceding digit is unchanged. If this digit is >5, it is eliminated and the preceding digit is increased by 1. If this digit equals 5, it is eliminated and the preceding digit is increased by 1.

Illustration of Rounding Numerical Values for Comparison with Requirements

Compendial Requirement	Unrounded Value (%)	Rounded Result (%)	Conforms
Assay limit ≥98.0%	97.96	98.0	Yes
	97.92	97.9	No
	97.95	98.0	Yes
Assay limit ≤101.5%	101.55	101.6	No
	101.46	101.5	Yes
	101.45	101.5	Yes
Limit test ≤0.02%	0.025	0.03	No
	0.015	0.02	Yes
	0.027	0.03	No
Limit test ≤3 ppm	0.00035	0.0004	No
	0.00025	0.0003	Yes
	0.00028	0.0003	Yes

16.3.3.2 Significant Figures

Great attention is paid to the use of significant figures in laboratory reports.

Electronic calculators are used to facilitate the computations associated with experimental numbers; even inexpensive calculators are designed so as to deal with, and display, numbers having 10 or more digits. This places the burden of accounting for the appropriate representation of numerical data on the investigator.

Everyday laboratory measurements limit the number of digits that represent reality.

The concept of significant figures seeks to limit the number of digits in a quantity to those that are justified based on its origin.

Two features play a role in the determination of significant figures:

a. Measurement
b. Computation using measured values

SOP No. QCS-016.00 Effective date: mm/dd/yyyy
Approved by:

16.3.3.2.1 Measurement

For a numerical quantity which is the result of a direct measurement, the number of decimal digits in that quantity is generally fixed by the measuring device. The number of significant digits can depend on the value of the measurement.

For experimentally determined quantities, the number of significant figures is not simply related to the intrinsic precision of a device. Consider the buret and the analytical balance.

The buret permits a volume to be read to ±0.02 mL (one-fifth of the smallest division on its scale). How many significant figures does such a quantity have?

If the volume is <1.00 mL, the number of significant figures in the volume is two (e.g., 0.57 mL). If the recorded volume is between 1.00 and 9.99 mL, the number of significant digits is three. If the volume is between 10.00 and 50.00 (the maximum capacity of the buret), the number of significant figures in the volume reading is 4, the maximum number of significant digits that a 50 mL buret is capable of producing.

Buret readings must always be reported to two decimal places, regardless of the volume reading. The number of significant figures in the reading clearly depends on the magnitude of the reading.

Next, consider the analytical balance.

The analytical balance is capable of measuring weights to ±0.0001 g.

The number of significant figures represented by this intrinsic precision varies with the total weight of an object. Nevertheless, weights determined on an analytical balance must always be reported to four decimal places, regardless of the weight. Again, the number of significant figures in the weight depends on the weight. For example,

For weights:

- <1 mg (0.0010 g), the balance produces only one significant figure (e.g., 0.0006 g = 0.6 mg).
- Between 1.0 and 9.9 mg (0.0099 g), the balance produces two significant figures (e.g., 0.0066 g = 6.6 mg).
- Between 10.0 and 99.9 mg (0.0999 g), the balance produces three significant figures (e.g., 0.0666 g = 66.6 mg).
- Between 100.0 and 999.9 mg (0.9999 g), the balance produces four significant figures (e.g., 0.6666 g = 666.6 mg).
- Between 1.0000 and 9.9999 g, the balance produces five significant figures (e.g., 6.6666 g).
- For weights between 10.0000 and 99.9999 g, the balance produces six significant figures (e.g., 66.6666 g).

16.3.3.2.2 Computation

For a computed quantity, the number of significant digits reflects the number of digits in the numbers from which it is computed.

The number of significant figures in a computed quantity may be the same as or different from those of the quantities from which it is computed. Consider a container that together with its contents weighs 35.2749 g. A sample of the contents is transferred out, producing a final weight of 35.2235 g. Both the initial and final weights are known with a precision of six significant figures. The weight of the sample is

$$\begin{array}{r} 35.2749 \text{ g} \\ -35.2235 \text{ g} \\ 0.0514 \text{ g} \end{array}$$

SOP No. QCS-016.00 Effective date: mm/dd/yyyy
Approved by:

The six significant figures in the initial and final weights produce a number (their difference) with only three significant figures (but still with four decimal digits. The zero after the decimal point, while important, is not a significant figure).

Suppose that the sample weight is now used in the calculation of a concentration by dividing it by a volume of 20.32 mL. Despite the four significant figures in the volume, the resulting concentration should be reported to at most three significant figures, namely 0.00253 g/mL. In general, the number of significant figures in a multiplication or division of quantities will be that of the quantity with the least number of significant figures.

16.3.3.2.3 How to Manage Zeros

Perhaps the most confusing aspect of significant figures is how to deal with the digit zero (0). That number is often used as a place holder to indicate the location of a decimal point. When used in this manner, zero is not a significant figure. The only times that 0 is significant is when one or more of them occur between nonzero digits or when one or more of them in a row is at the end of a string of nonzero digits.

Consider the following examples:

The 0 is used only to highlight the occurrence of a decimal point following it. It is not a significant figure.

0.0654 In this case, neither 0 is significant. They are not preceded by any nonzero digit

0.605 The 0 is again not significant, but the 0, since it occurs between two nonzero digits, is significant

0.6005 The 0 is again not significant, but both the 0's are significant, since they occur between two nonzero digits.

10.654 This time, the 0 to the left of the decimal point is significant. It occurs between two nonzero digits.

10.0654 In this number, both the zeros are significant digits since there are nonzero digits both before the first and after the second.

Now consider a number that is the result of an experimental measurement. Suppose the number is reported as 3.16.

We can assume that the investigator was certain that the result was <3.17 and >3.15. With the usual rules for rounding, we can indeed assume that the result was <3.165 and >3.155.

Suppose instead, one reports the number as

3.160 What shall we make of the terminal 0? The fact that the number is written as is an indication that the investigator was certain that the result was <3.161 and >3.159. (With rounding, <3.1605 and >3.1595.) A substantial decrease in the uncertainty of the result and an increase in the precision of the measurement over the previous representation.

If the number is reported as

3.1600 This is an indication that the investigator was certain that the number was 3.1599. (One should be able to provide the limits assuming rounding rules.) An additional decrease in the uncertainty. These zeros are certainly significant.

If one understands the above, one should be able to determine which zeros are significant in the following number: 0.0010600.

SOP No. QCS-016.00 Effective date: mm/dd/yyyy
Approved by:

16.3.3.2.4 Integers: A Special Case

For quantities that are restricted to occurring in discrete, indivisible units, the number of significant figures is infinite. We generally assume that integers always have as many significant figures as are necessary when used in a calculation. For example, in the calculation of the perimeter of a square (four times the length of a side), we can assume that the number 4 used in the calculation has as many significant figures as the length measurement. One exception to this rule is Avogadro's number which, while an integer, has not been known to the requisite 23 significant figures.

The astute student will recognize that numbers such as pi, while not an integer, can be represented to as many significant figures as may be required in any given computation (e.g., the calculation of the volume of a sphere from its radius, using $V = (4/3) \times \text{pi} \times r^3$). Indeed, in this formula, the 4, the two 3s and the pi are all known to an infinite number of significant figures. All of the uncertainty in the volume V is due only to the uncertainty in the radius r. The number of significant figures in V is fixed by those in r.

16.3.3.2.5 Significant Figures and Scientific Notation

Scientific notation (i.e., writing numbers in the form y.yyy × 10*n*) simplifies the use of significant figures considerably. The issue of using zeros as place holders for decimal points can be made to disappear. The multiplier (y.yyy) has exactly the number of significant figures as are justified. This may involve a number of trailing zeros. For example, 6.02214×10^{23} has six significant figures. While scientific notation goes a long way toward solving the significant figure problem, it is not always easy to follow, particularly when additions and subtractions are involved. Consider, for example, the following two buret readings whose difference is the net volume.

Final volume 3.105×10^{1} mL
Initial volume 3.2×10^{-1} mL

In normal numeric representation, that subtraction is represented as follows:

Final volume 31.05 mL
Initial volume 0.32 mL

As the last example shows, in the laboratory course, scientific notation will not always be the best way to handle simple computations—particularly the addition and subtraction of numbers.

16.4 REASON FOR REVISION

First time issued for your company, affiliates, and contract manufacturers.

QCS-017

YOUR COMPANY STANDARD OPERATING PROCEDURE

SOP No. QCS-017.00 Effective date: mm/dd/yyyy
Approved by:

TITLE: **Analytical Methods Transfer for New Products**

AUTHOR: ________________________________

Name/Title/Department

____

Signature/Date

CHECKED BY: ________________________________

Name/Title/Department

____

Signature/Date

APPROVED BY: ________________________________

Name/Title/Department

____

Signature/Date

REVISIONS:

No.	Section	Pages	Initials/Date

SOP No. QCS-017.00 Effective date: mm/dd/yyyy
Approved by:

Subject: Analytical Methods Transfer for New Products

17.1 PURPOSE

The purpose of this procedure is to outline the process and requirements for transfer of validated analytical methods from third-party laboratories to the other manufacturing site.

17.2 SCOPE

This procedure must be followed for any transfer of validated chemical or physical or analytical methods in support of a new product transfer. This will also apply for the transfer of new analytical methods in support of an existing product.

17.3 RESPONSIBILITY

It is the responsibility of the QC analysts to follow the procedure. The QC manager is responsible for SOP compliance.

17.4 DEFINITIONS

1. *Chemical or physical analytical method:* Any analytical method that involves physical or chemical measurements including but not limited to, chromatographic, titrimetric, and spectrophotometric methods and excluding biological and microbiological methods of analysis. Limit tests are not included.
2. *Transfer of validated chemical or physical analytical methods:* The transfer of a method from the contract manufacturing laboratory to any other company's laboratories.
3. *Project leader:* The project leader is the person in the contract manufacturing laboratory assigned to define which methods require transfer, and executes and monitors the method transfer process.
4. *QC analyst:* The QC analyst is the person assigned at any one of the other company's laboratories, in coordination with the project leader, to properly execute the experiment involved in the method transfer, and prepare a report summarizing the results of the study.
5. *Project supervisors:* The project supervisor is the person assigned the overall responsibility for the proper and timely execution, and reporting of, as well as approval of the results of the transfer.

17.5 PROCEDURE

1. The project leader, in coordination with the QC manager, would review the Method and Validation Report to arrive at a decision with justification for a recommendation for a formal transfer of the method. Such recommendation should be approved by the analytical manager or designate. The United States Pharmacopeia (USP) test methods will not require a formal transfer. Also excluded are simple in-house developed test methods which have characteristics similar to the existing QC test methods for which the QC laboratory has demonstrated expertise.

SOP No. QCS-017.00 Effective date: mm/dd/yyyy
Approved by:

2. The project leader, project supervisors, and QC analysts must schedule a method transfer review meeting prior to initiation of the formal transfer study. At this meeting a review of the method development and validation and test method SOP should be conducted. Where necessary, a trial run should be performed to verify proper operation of equipment and understanding of the test method SOP prior to initiation of the study. The project leader from contract manufacturing laboratory will remain actively involved to resolve any issues related to the execution of the test method SOP.
3. A representative sample will be obtained by the project leader and provided to the QC analyst. The project leader should ensure that proper identification/documentation of the sample is received by the QC analyst before the experiment is performed. The QC analyst must also ensure that the proper documentation is received.
4. Other necessary materials, such as the validation report, a clear and thorough SOP, reference standard, reagent-type information, and so on must also be documented and furnished by the project leader to the QC analyst.
5. For those methods which require a formal transfer study, a precision run (repeatability) will be performed at the plant QC laboratory. Six independent test preparations from a homogenous sample will be performed per the approved test method SOP.
6. Results will be reviewed, and approved if acceptable, by the QC manager.
7. The QC manager will prepare a concise report. Review and approval will be obtained from the project supervisors at each site.
8. If the acceptance criteria are not met, an investigation will be conducted. If the source of the error is identified, corrective action is implemented and the precision run is repeated. If the source of error is not identified, two more precision runs are performed for further evaluation. If a conclusion still cannot be arrived at, an investigation and course of action will be developed and documented.

17.6 ACCEPTANCE CRITERIA

1. System suitability limits must be met.
2. The % RSD of the six preparations meet the repeatability criterion established in the Method Validation Report and the average of the six results meet the regulatory specifications.
3. For chromatographic methods, chromatograms are comparable to those obtained in the Method Validation Report.

17.7 DOCUMENTATION

Each step in the method transfer report will be documented. Documentation required in the SOP or monograph being tested and any internal SOP on documentation will be followed.

The results of the transfer study will be documented on the attached report form (Attachment I).

17.8 REASON FOR REVISION

First time issued for your company, affiliates, and contract manufacturers.

SOP No. QCS-017.00 Effective date: mm/dd/yyyy
Approved by:

17.9 ATTACHMENT FORM

ABC Pharmaceutical Company

SOP No.:
Issued on:
Revision No.:
Initiator name:

Attachment I

ANALYTICAL METHODS TRANSFER

VALIDATION FINAL REPORT

Page 1 of 3

Validation type: Finished product	☐	Raw materials	☐
Assay	☐	Assay	☐
Impurities	☐	Residual solvents	☐
		Organic volatile impurities	☐

Product/Raw material name: ____________________________ Lot # ______________________

Product codes: ____________________ ____________________

____________________ ____________________

____________________ ____________________

Drug product concentration: _______________ Raw material concentration: _______________

1.0	Assay design key parameters	
1.1	See attached method validation report	# ________________________
1.2	See attached governing method SOP	# ________________________

2.0	Assay method key data section			
2.1	System suitability			
	Parameter	**Limits**	**Results**	**Comments**
	RSD of replicate injection			
	Tailing factor			
	Resolution			
	____________________________________			
2.3	Data referenced location	Laboratory notebook #		Pages
			Storage location:	

SOP No. QCS-017.00 Effective date: mm/dd/yyyy
Approved by:

ANALYTICAL METHODS TRANSFER VALIDATION FINAL REPORT

2.4	Assay method—Sample replicate data section		
	Sample Preparation	**Average %**	**Comments**
	1		
	2		
	3		
	4		
	5		
	6		
	Average % RSD limit	________ ________ ________	
2.5	Data referenced location	Laboratory Notebook #	Pages:
		Storage location	

3.0	Impurities/Residual solvent method key data section			
3.1	System suitability			
	Parameter	**Limits**	**Results**	**Comments**
	RSD of replicate injection			
	Tailing factor			
	Resolution			
	________			

SOP No. QCS-017.00 Effective date: mm/dd/yyyy
Approved by:

ANALYTICAL METHODS TRANSFER

VALIDATION FINAL REPORT

3.2	Impurities/Residual solvent sample replicate					
	Sample Preparation	**Component 1**	**Component 2**	**Component 3**	**Component 4**	**Component 5**
	1					
	2					
	3					
	4					
	5					
	6					
	Average % RSD limit	______ ______ ______	______ ______ ______	______ ______ ______	______ ______ ______	______ ______ ______
3.3	Data referenced location			Laboratory Notebook #		Pages
				Storage location:		

4.0	Conclusion
4.1	The method referenced in this report conforms/does not conform to the predetermined limits specified in the SOP.

Data summarized by: ______________________ Date: __________

Testing conducted by: ______________________ Date: __________

Study director transfer site: ______________________ Date: __________

Transfer site lab management: ______________________ Date: __________

Attachments:
Method Validation Report
Governing Standard Test Method

QCS-018

YOUR COMPANY STANDARD OPERATING PROCEDURE

SOP No. QCS-018.00 Effective date: mm/dd/yyyy
Approved by:

TITLE: **Handling Out of Specification Results**

AUTHOR: ______________________________

Name/Title/Department

____

Signature/Date

CHECKED BY: ______________________________

Name/Title/Department

____

Signature/Date

APPROVED BY: ______________________________

Name/Title/Department

____

Signature/Date

REVISIONS:

No.	Section	Pages	Initials/Date

SOP No. QCS-018.00 Effective date: mm/dd/yyyy
Approved by:

18.1 PURPOSE

This SOP describes the procedure of reporting and investigating out-of-specification (OOS) or atypical results. It applies to quantitative and qualitative tests generated in the QC laboratory. This procedure also describes the process for retesting following the laboratory investigation into a laboratory error.

18.2 SCOPE

This procedure applies to all test results generated at the QC laboratory that do not conform to specifications. This procedure is applicable to OOS investigation of raw material, in-process, bulk, final container, and stability (including ongoing and accelerated studies) test results. The purpose of the investigation is to determine that the results are correct and a laboratory error has not occurred.

As appropriate, the procedure may also be used in the investigation of suspect or out-of-trend (OOT) test results that are within specification.

REFERENCES

1. Investigating Out of Specification (OOS) Test Results for Pharmaceutical Production, FDA—CDER Draft Guidance, 1998.
2. USA 21 CFR, Parts 210 and 211 "Current Good Manufacturing Practices."

18.3 DEFINITIONS

Assignable cause: An event supported by documented evidence, which has resulted in an erroneous test value or an uncompleted set of data.

Control sample: Samples with known test value, such as a retain sample of previously released batches.

Error: The difference between observation or calculated value and the true value due to variations in measurements, calculations, or observations of a quantity due to mistakes. Unintentional deviation from a standard or specification.

Investigation: A systematic examination and inquiry of an incident, condition, or situation.

Out-of-specification (OOS): Any individual test value that does not conform to specification.

Out-of-trend (OOT): Any individual test value that is within specification but is scientifically questionable or does not follow expected trends.

Resample: A new representative of the lot or batch obtained using the normal sampling procedure, or a sample collected from original containers and same portion.

Retest: A repeat of a test using the original laboratory sample.

Suspect test result: An unexpected test result based on historical analytical trends and/or current material action limits.

Unexpected result: Any result that is out of specification or out of trend.

18.4 PROCEDURE

1. General
    a. If there is sufficient evidence, the product/material may be rejected at any time during the OOS investigation (e.g., presence of foreign material in the sample or the material is not homogeneous).

SOP No. QCS-018.00 Effective date: mm/dd/yyyy
Approved by:

b. Whenever retesting is called for in this procedure, retesting must be conducted on the original laboratory sample. A resample can be done where
   i. Inappropriate sample has been identified as the assignable cause of unexpected result.
   ii. There is insufficient original laboratory sample to complete all the required testing.
   iii. A process investigation is required.
c. If an error is detected during the course of any analysis, the analyst/technician must not knowingly continue the testing. Where no result on any test sample has been generated, the test run may be invalidated with documented rational (on the laboratory notebook), and with supervisor/manager approval, the test may be restarted.

18.5 RESPONSIBILITY

18.5.1 QC Analyst

1. It is the responsibility of all QC analysts to notify their supervisors of any nonconformity, OOS, or suspect test results.
2. Reporting initial test results: (see Form 01)
   a. The analyst must notify the responsible supervisor of an OOS result generated.
   b. The analyst must retain all test samples, standard solutions, equipment and instruments used for testing, and glassware, including pipettes and syringes, until the obtained data has been analyzed. These may be used for further investigation.

Note: Investigation of test result failures is essential in confirming whether an OOS result can be attributed to error or whether the result represents the actual condition of the material or process being evaluated.

   c. The analyst must start the OOS lab investigation report (see Form 01) by filling out the header information, including product/material name, batch/lot number, analyst name, and date.
   d. The analyst must describe the unexpected result in the section provided and state the test limits or acceptance criteria, the laboratory notebook, and page reference.

18.5.2 Preliminary Lab Investigation: (See Form 01)

18.5.2.1 QC Analyst and Supervisor

1. It is the responsibility of the QC supervisor to inform the QC laboratory manager and perform preliminary investigation.
2. The analyst and lab supervisor must conduct a preliminary investigation to determine if an assignable cause for the unexpected result exists. The investigation may include but is not limited to any or all of the following areas.

18.5.2.1.1 Raw Data Review

- Was the calculation correct?
- Was the work done accurately?
- Comparison of results to certificate of analysis (CoA), historical data, or trends.

SOP No. QCS-018.00 Effective date: mm/dd/yyyy
Approved by:

18.5.2.1.2 Sample Consideration

- Was the correct sample tested?
- Is there any evidence of sample contamination?
- Is there any evidence of improper sample appearance, storage, handling, or damage?

18.5.2.1.3 Equipment

- Was the correct equipment used?
- Were the settings correct?
- Is there any indication of equipment malfunction?
- Is the equipment calibration/maintenance current?

18.5.2.1.4 Test Performance

- Was correct test method used?
- Was raw data properly documented?
- Was the analyst trained on the technique?

18.5.2.1.5 Reagents, Standards, and Sample Preparation

- Were correct standards and reagents used?
- Were there any expired reagents, chemicals, or standards?
- Was the test sample homogeneous?
- Were the test solutions correctly prepared?

18.5.2.1.6 Other Error

Additional questions/information pertaining to specific test method, technique, or apparatus should be recoded in this section by the analyst and the supervisor as part of determining the cause of OOS test results.

18.5.3 Preliminary Lab Investigation Conclusion: (See Form 01)

18.5.3.1 QC Supervisor

1. When the initial investigation demonstrates that the OOS result is attributable to *determinate* (i.e., *assignable*) analytical identifiable *error*, the supervisor must use the check box to indicate that an assignable cause has been found.
2. The lab supervisor must then provide a brief description of the cause of the analytical error, evaluate the possible corrective actions, and collaborate with QA department in implementing them.
3. The supervisor must perform an impact assessment to determine the extent of retesting required:
    a. Determine if the error has compromised the results of other tests.
    b. Evaluate the need for retest of other materials/products in order to confirm the validity of their initial results.
4. The supervisor must instruct the original analyst to conduct a retest.

18.5.3.2 QC Analyst

5. The analyst must repeat the test as per the original method, on the initial aliquot from the original sample.

SOP No. QCS-018.00 Effective date: mm/dd/yyyy
Approved by:

Note:

- If the initial aliquot from the original sample is not sufficient for retest, a second aliquot from the original sample is used.
- In the event that the original sample has been consumed during analysis, a second sample (new sampling) can be used.
- When the original test samples are found to be the cause of OOS results, a new set of samples must be used for retest.

6. Analyst must record and report the retest results in the laboratory notebook.

18.5.3.3 QC Supervisor

7. If all repeat test values are within the specification:
   a. The supervisor must invalidate the initial test results and report the repeat test values as final.
   b. The supervisor must sign off the OOS report and strike out the other parts of the OOS form (see Form 01).
   c. The supervisor must then forward the OOS form to the QC manager, director, and QA director for approval.
8. If the repeat test values are out of specification (OOS), the supervisor must investigate the cause of failure according to the preliminary laboratory investigation section.
9. If the preliminary investigation of the OOS test result demonstrates that it is not attributable to *indeterminate* (i.e., *nonassignable* or *unknown*) error, the supervisor must use the check box to indicate that a nonassignable cause has been found.

Note: Statistical tests have their own retest processes, following the compendia guidelines. For these exceptions, failure to meet the criteria of any stage prior to the last stage is not a test failure and does not require the OOS investigation report.

10. Where the initial aliquot from the original sample (i.e., working test sample) is sufficient, a repeat test must be performed to confirm initial OOS test results under identical condition as the initial test.
11. The result of nonassignable preliminary OOS investigation must be reported to the QC manager and director and QA department immediately.

18.5.4 OOS Retest Plan: (See Form 01)

18.5.4.1 QC Manager

1. It is the responsibility of the QC manager to notify the QA department when the preliminary investigation is inconclusive.

18.5.4.2 QA Specialist

2. The QA department is responsible for reviewing, following up, and approving all investigations of OOS test results, decision for final disposition of material, and issuance of the final report.

18.5.4.3 QC Manager and QA

3. The QC manager and QA must lead the investigation from this point onward when the result of the preliminary investigation is inconclusive.
4. The QC manager and QA must decide on retest plan.

SOP No. QCS-018.00 Effective date: mm/dd/yyyy
Approved by:

5. The OOS investigation form must be completed within 30 working days and forwarded to the QA department for review and approval.

Note: Due to the nature of some biological tests, the OOS completion may take longer than 30 working days. Under these circumstances, efforts should be made to complete the investigation in a timely manner.

6. The retest strategy and the analyst who performs the testing must be determined by the QC laboratory manager.
7. The retest plan and number of repeat tests should be specified in advance prior to any testing.

18.5.4.4 QC Analyst

8. The laboratory analyst must repeat the test for at least three times using fresh preparation for each repeat from the original laboratory sample for each repeat.
9. Results must be reported in the investigation of OOS.
10. OOS repeat results: (see Form 01)

If	And	Then
Any repeat test value is OOS	No assignable cause is identified	The batch fails the test
	An assignable cause is identified	Follow corrective actions for assignable cause section
Each and every repeat test value is within the specification		The batch passes the test Report the average of all repeat test values as the final result

18.5.5 Final Review, Corrective Actions, and Conclusions: (See Form 01)

18.5.5.1 QC Supervisor

The lab supervisor must summarize all repeat test results in the "retest reporting" section of the OOS form and attach copies of any supporting information or data summaries to the OOS report.

18.5.5.2 QC Manager

1. The QC laboratory manager must provide the laboratory's conclusion to the OOS result in the report.

If the Batch	Then
Passes the test	The QC manager should check "analytical property meets specification."
Fails the test and a disposition decision is required (including stability testing failures when the product has not expired)	The QC manager should check "analytical property does not meet specification." The QC manager should ensure the quality assurance is notified. The QC manager should provide a brief summary of the conclusion of the investigation.
Fails the tests and samples are within their shelf lives (for stability samples only)	The QC manager should inform the QA. The QC manager should inform registration and export director immediately for the determination of the appropriate reporting of the OOS test results on the product in the market.

SOP No. QCS-018.00 Effective date: mm/dd/yyyy
Approved by:

2. QC laboratory manager must fill in the "comment" section including an impact analysis, further action, and propose the disposition of affected batches.
3. QC laboratory manager should forward the original form to the QA department for final approval.

18.5.5.3 QA Director

The decision for the rejection or the release of the product impacted by the OOS result is determined by the QA director.

18.5.5.4 Regulatory Affairs Manager/Director

Where applicable, within 3 working days, the director—regulatory affairs must inform the related regulatory agencies and customers of the failing marketed batches and collaborate with them in disposition of these batches from the marketplace.

18.6 ATTACHMENTS

- OOS Investigation: Form 01
- OOS Investigation Process Flow Chart: Attachment I

18.7 REASON FOR REVISION

First time issued for the ABC Pharmaceutical Company.

SOP No. QCS-018.00 Effective date: mm/dd/yyyy
Approved by:

18.8 ATTACHMENT FORMS

ABC PHARMACEUTICAL COMPANY

LEVEL 3
Page 1 of 4
Attachment No.:
Issued on:
Revision No.:
Initiator name:

OUT-OF-SPECIFICATION RESULTS

FORM 01

Originator: ____________________ **Title:** ______________________ **Date:** ________________

INVESTIGATION AND DISPOSITION FORM

Product: ________________	Code: __________________	Batch No.: ________________
Material: ________________	Code: __________________	Lot No.: ___________________
Test: ________________	Method No.: ____________	Acceptance Criteria: ________

Problem: (Results not meeting defined acceptance criteria) OOS description:

Actual Results:

A. Preliminary lab investigation checklist	**Yes**	**No**	**N/A**
• Analyst trained in the procedure?	☐	☐	☐
• Instrument failure (correct equipment or instrument used)	☐	☐	☐
• Reagent problem (reagent/standard/media expired)	☐	☐	☐
• Nonrobust method (method validation)	☐	☐	☐
• Was the correct method used?	☐	☐	☐
• Were there any deviations from the method?	☐	☐	☐
• Were there any arithmetical errors in calculations? (calculation data correct)	☐	☐	☐
• Was the equipment functioning correctly?	☐	☐	☐
• Was the equipment calibrated?	☐	☐	☐
• Were reagents, prepared samples, and standards correctly prepared?	☐	☐	☐
• Were system suitability requirements defined in the method met?	☐	☐	☐
• Were acceptance criteria defined in the method for precision of replication met?	☐	☐	☐
• Were there any obvious physical problems with the laboratory sample?	☐	☐	☐

SOP No. QCS-018.00 Effective date: mm/dd/yyyy
Approved by:

LEVEL 3
Page 2 of 4
Attachment No.:
Issued on:
Revision No.:
Initiator name:

A. Preliminary lab investigation checklist (continued)	Yes	No	N/A
• Were positive or negative controls satisfactory?	☐	☐	☐
• Equipment sterile	☐	☐	☐
• Correct glassware used	☐	☐	☐
• Pipettes calibrated	☐	☐	☐
• Dilution made correctly	☐	☐	☐
• Test tube labeled correctly	☐	☐	☐
• Proper storage conditions	☐	☐	☐
• Incubation conditions adequate	☐	☐	☐
• Sterilization complete	☐	☐	☐
• Gowning adequate	☐	☐	☐

Other error/s (please specify any other parameter that may contribute to OOS results):

____

____

____

____

____

____

____

Cause: • **Assignable** • **Nonassignable**

SOP No. QCS-018.00 Effective date: mm/dd/yyyy
Approved by:

LEVEL 3
Page 3 of 4
Attachment No.:
Issued on:
Revision No.:
Initiator name:

B. Expanded investigation:

- Review information available about the material under test
- The manufacturing process
- History of the batches
- Results of in-process testing
- Results of other tests conducted
- Supplier certificate of analysis
- Deficiencies deviation identified during review
- Actual and probable cause

SOP No. QCS-018.00 Effective date: mm/dd/yyyy
Approved by:

LEVEL 3
Page 4 of 4
Attachment No.:
Issued on:
Revision No.:
Initiator name:

OUT-OF-SPECIFICATION RESULTS

INVESTIGATION AND DISPOSITION FORM

C. Conclusion and corrective action:

____
____
____
____
____
____
____

QC manager

D. Preventive measure:

____
____
____
____
____
____
____

QC manager

E. Repeat/retest results and disposition:

____
____
____
____
____
____
____

QC manager

CC: BATCH FILE

QCS-019

YOUR COMPANY STANDARD OPERATING PROCEDURE

SOP No. QCS-019.00 Effective date: mm/dd/yyyy
Approved by:

TITLE: **Batch Auditing Procedure**

AUTHOR: ______________________________

Name/Title/Department

____

Signature/Date

CHECKED BY: ______________________________

Name/Title/Department

____

Signature/Date

APPROVED BY: ______________________________

Name/Title/Department

____

Signature/Date

REVISIONS:

No.	Section	Pages	Initials/Date

SOP No. QCS-019.00 Effective date: mm/dd/yyyy
Approved by:

19.1 PROCEDURE

To provide a written procedure for the batch documents review using a standard checklist by weighing area, manufacturing, in-process control (bulk), packaging, in-process control (packaging), QC laboratory and finally QA batch documents checklist to ensure completeness of records.

19.2 SCOPE

The SOP is applicable to all product batches, manufactured in the ABC Pharmaceutical Company. All related checklists will be attached along with the master manufacturing formula, master packaging instructions, manufacturing in-process, and packaging. For the US market intended products (ANDA submissions) a final batch audit will be conducted prior to the QC release. The findings for all parenteral products will be recorded in Annexure VII.

19.3 RESPONSIBILITY

1. This is the responsibility of the corresponding departmental supervisors to implement the procedure.
2. The QA batch auditor is responsible for the audit of the final batch documents.
3. The department managers and directors are responsible for SOP compliance.

19.4 PROCEDURE

19.4.1 Weighing Area

At the end of weighing, the area supervisor will audit the following weighing records:

- Batch manufacturing record for the presence, correctness, and completeness of information regarding the charge quantity and if the purchase order number in the specified column is duly signed.
- Raw materials requisition for the presence, correctness, and completeness of the deliverd quantity, and the signatures of the dispensing, issuing, and receiving personnels.
- Raw material weighing records with the printouts of the weighed raw material. Weighing area clearance complete with the signatures of the weighing area in-charge.

These documents will be forwarded to the manufacturing area supervisor for further processing. For audit use Attachment I.

19.4.2 Bulk Product Manufacturing

After the completion of bulk manufacturing, the area supervisor will audit the following documents.

1. Batch manufacturing record, check that
   i. All the information regarding the identity of the batch is correct
   ii. Record is duly signed off by the manufacturing area supervisor
   iii. Information related to machine cleaning is complete with the signature of the worker
   iv. Yield calculation is correct
   v. Temperature and timings and other entries mentioned are as per the master formula

SOP No. QCS-019.00 Effective date: mm/dd/yyyy
Approved by:

2. Batch manufacturing records
   - Information required for the identity of the batch are correct
   - Record is duly signed by the manufacturing supervisor
   - Information related to the machine cleaning and sterilization is complete with the print outs and signatures
   - Yield calculation is correct
   - All the entries related to temperature, humidity, and others are recorded and are within the range
   - Weight variation record of the tablets compression and capsules filling generated by machine operators is present and within the limits
   - Ampoules/vials filling record are present and are within the limits
3. Batch manufacturing record for coating is present and all the information is correct
4. Record for the presence of any deviation from the master manufacturing formula, if yes, then any corrective action is taken, the process change request is raised and made part of the batch documents
5. Final batch yield and yield at various in-process stages is correct
6. Rejection verification at various stages of processing and at the final stage

19.4.2.1 Injectables

The following checks are carried out:

- Temperature and pressure differential records are available
- Sterilization record for machine parts, filters, and gowns
- Ampoule/vials washing and sterilization records
- Filter integrity test report
- Area/equipment clearance (visual inspection)
- Visual inspection request
- Print out for leak test performed
- Print out for terminal sterilization (if applicable)

19.4.2.2 Lyophilized Products

- Lyophilization cycle chart

19.4.2.3 Powder-Filled Vials

- Seals/stoppers sterilization records
- Cramper sterilization record

For audit checklist, refer to Attachment II.

19.4.3 In-Process Control (Bulk Product)

The QA inspector (bulk manufacturing area) is responsible to audit the following documents at the end of the process before sending the samples and documents to the QC laboratory. The person at the quality function unit will be an authorized one from the QA department.

19.4.3.1 Weighing Area Records

- Batch manufacturing record No. 1
- Raw materials weighing record with print out
- Weighing area clearance checklist

SOP No. QCS-019.00 Effective date: mm/dd/yyyy
Approved by:

19.4.3.2 Production Area Documents

- Batch manufacturing record
- Master manufacturing formula
- Bill of materials
- Sanitization record of the area (suspension)

19.4.3.3 QA In-Process

- Sampling report (manufacturing side)
- QA in-process records
    - Weight variation format (tablets, capsules, suppositories, and parenterals)
    - Inspection start-up checklist with release label
    - Visual inspection for parenterals
    - Weighing area checklist signed by the QAI

19.4.3.4 Any Other Documents

- Process change request
- Memo (sorting or any other)
- Waiver document

For audit checklist, refer to Attachment III.

Note: Before starting the packaging operation, the packaging supervisor and the QAI must ensure that the batch was released for packaging and that the bulk release sheet was included as per the risk packaging approval SOP.

19.4.4 Packaging (Finished Product)

At the end of the packaging operations the packaging area supervisors are responsible to carry out the audit of the following documents and verify the

- Batch packaging record for the presence, correctness of all the entries, and the correctness of manufacturing and expiry dates as per the shelf life
- Packaging report and the release and finished product report
- Packaging materials requisition (computer copy)
- Batch final yield and at various in-process stages are correct
- Rejection verification at various stages of processing and at final stage
- Reconciliation of labels and boxes done by the packaging department
- Area/equipment line clearance for packaging operation
- Master packaging instructions
- Output sheet
- Repacking addendum (if applicable)
- Store requisition for the repacking material
- Batch packaging record for
    - Blister packing
    - Suppositories filling and packing
    - Liquid filling and packing
    - Powder filling and packing
    - Bottle packing (tablet/capsule)

SOP No. QCS-019.00 Effective date: mm/dd/yyyy
Approved by:

- Bottle torque (if applicable)
- Sachet filling and packing
- Sachet seal test

For audit checklist, refer to Attachment IV.

19.4.5 In-Process Control (Finished Products)

The finished product packaging area QAIs are responsible to audit the following documents at the end of packaging operations. The person at the quality function unit will be an authorized one from the QA department.

19.4.6 Packaging Area Documents

Batch packaging record

- Hand packing
- Filling and packing
- Line clearance format
- First signed coded commodities
- Reconciliation form (labels/cartons)
- Packaging report and the release of finished product (for different consignees)
- Packaging material requisition
- Time sheet

19.4.6.1 QA In-Process

- Sampling report (manufacturing side)
- QA in-process record
  - QA filling chart (cream, ointment, powder filling, suspensions, syrups, drops, blister or strips (tablet/capsule), sachets)
  - QA final packaging records (for different consignees)
  - Inspection start-up checklist with release label

19.4.6.2 Any Other Documents

- Process change request
- Memo (sorting or any other)
- Repacking request/requisition

For audit checklist, refer to Attachment V.

19.4.6.3 Finished Product Quality Control

- The QC officer will be responsible for checking the availability of completed Attachments I through V. In addition to this, the QC officer will also check that
  - Test results as reported by the QC analyst are within the specification limits and that there is no nonconformance in any of the tests
  - The information related to the batch identity is available and correct
  - For all manufacturing processes
    - Analytical report (bulk product)
    - Microbiological reports (when applicable)

SOP No. QCS-019.00 Effective date: mm/dd/yyyy
Approved by:

- For parenteral products
- Sterility report (for parenteral products)
- LAL test report
- Air monitoring record of the filling room
- Personnel hygiene monitoring report
- Surface monitoring report
- Water testing report
- Gloves print (operator) report
- Machines swabs report

- The deviation or remarks noted in the batch documents or any deviation related to the batch manufacturing or packaging shall be rectified prior to the final release.

Final certification of release for each consignee will be issued by the QC manager/designate. This will be done by issuing a QC number for each consignee on the packaging report and release of finished product. Separate Certificate of Release will not be issued.

19.4.7 Quality Assurance Audit Documents

Annexures VII and VIII are for final batch audit of parenteral products and rest of the products, respectively. These checklists cover the whole batch documents for their presence, correctness, and completeness.

19.4.7.1 Documentation

- Attachment I (weighing area)
- Attachment II (bulk product manufacturing)
- Attachment III (in-process control—bulk)
- Attachment IV (packaging finished product)
- Attachment V (in-process control—finished products)
- Attachment VI (finished products quality control)
- Annexure VII (QA final batch documents checklist for injectables)
- Annexure VIII (checklist for final QA batch document audit)

19.5 REASON FOR REVISION

First time issued for the ABC Pharmaceutical Company.

19.6 ATTACHMENT FORMS

Attachment I
SOP No.: QCS:
Issued on:
Revision No.:
Initiator name:

DOCUMENTS COMPLIANCE DECLARATION

Weighing & Dispensing Raw Materials

Product : ______________________________

Code No. : ______________________________

Batch No. : ______________________________

This is to confirm that the batch documents sent have been verified against the checklist for the completeness, correctness, clarity, accuracy, and consistency as per the requirement of cGMP and ABC Pharmaceutical Company's standards.

Compliance check by: ______________________

Signature: _____________

Date: _________________

Attachment I
SOP No.: QCS:
Issued on:
Revision No.:
Initiator name:

CHECKLIST FOR THE BATCH DOCUMENT AUDIT

(Weighing Area)

1. Batch manufacturing record for the presence, correctness, and completeness of information regarding the charge quantity and the purchase order number in the specified columns duly signed. ☐
2. Raw materials requisition for the presence, correctness, and completeness of the delivered quantity, QC number and the signatures of the dispensing, issuing, receiving personnels. ☐
3. Raw material weighing records with the print outs of the weighed raw material. ☐
4. Weighing area clearance complete with the signatures of the weighing area in-charge. The area/booth cleaning status label "CLEANED" to be made part of the batch record. ☐

Any other documents

____

____

____

____

Reviewed by: ____________________

Applicable & available = ✓
Applicable not available = ?
Not applicable = ×

Attachment II
SOP No.: QCS:
Issued on:
Revision No.:
Initiator name:

DOCUMENTS COMPLIANCE DECLARATION

Manufacturing

Product : ______________________________

Code No. : ______________________________

Batch No. : ______________________________

Stage : ______________________________

This is to confirm that the batch documents sent have been verified against the checklist for the completeness, correctness, clarity, accuracy, and consistency as per the requirement of cGMP and ABC Pharmaceutical company's standards.

Compliance check by: ______________________

Signature: ____________

Date: ________________

Attachment II
SOP No.:
Issued on:
Revision No.:
Initiator name:

CHECKLIST FOR THE BATCH DOCUMENT AUDIT

(Manufacturing/Production Area Documents)

1. Batch manufacturing record
   On the batch manufacturing record check that
   - i. All the information regarding the identity of the batch is correct. ☐
   - ii. Record is duly signed by the manufacturing supervisor. ☐
   - iii. Information related to machine cleaning is complete with the signatures of the worker. The equipment cleaning status label "CLEANED" to be made part of the batch record. ☐
   - iv. Yield calculation is correct. ☐
   - v. Temperature and timings and other entries mentioned are as per the MFM. ☐
2. Batch Manufacturing record
   Compression (weight variation chart)
   Capsule filling (weight variation chart)
   Ampoule/vials filling (volume variation chart)
   - i. Information required for the identity of the batch is correct. ☐
   - ii. Record is duly signed by the manufacturing supervisor. ☐
   - iii. Information related to the machine cleaning sterilization is complete with the print outs and signatures. The equipment cleaning status label "CLEANED" to be made part of the batch record. ☐
   - iv. Yield calculation is correct. ☐
   - v. All the entries related to temperature, humidity, and other information are recorded and within the range. ☐
   - vi. Weight variation record of the compression and capsule filling generated by machine operators is present and within the limits. ☐
3. Batch manufacturing record for coating is present and all the information is correct. ☐
4. In case there was a deviation from the master formula, any corrective action taken, process change request raised and made part of the batch documents? ☐

Attachment II
SOP No.:
Issued on:
Revision No.:
Initiator name:

5. Batch yield at final and at various in-process stages is correct. ☐
6. Rejection verification of various stages of processing and at final stage. ☐

Injectables

7. Temperature and pressure differential records. ☐
8. Sterilization record for machine parts, filters, and gowns. ☐
9. Ampoule/vials washing and sterilization records. ☐
10. Filter integrity test report. ☐
11. Area/equipment clearance (visual inspection). ☐
12. Visual inspection request. ☐
13. Print out for leak test performed. ☐
14. Print out for terminal sterilization (if applicable). ☐

Lyophilized products

15. Lyophilization cycle chart. ☐

Powder filled vials

16. Seals/stoppers sterilization records. ☐
17. Cramper sterilization record. ☐

Any other documents

____

____

____

____

Reviewed by: ____________________

Applicable & available = ✓
Applicable not available = ?
Not applicable = ×

LEVEL 4
Attachment III
SOP No.:
Issued on:
Revision No.:
Initiator name:

CHECKLIST FOR BATCH DOCUMENTS ALONG WITH SAMPLES (BULK PRODUCT)

Product: ____________________ **Mfg.:** ____________________

Code No.: ____________________ **Expiry:** ____________________

Batch No.: ____________________

Stage: Bulk

Weighing Area Records

1. Batch manufacturing record ☐
2. Raw materials weighing record with print out ☐
3. Weighing area clearance check list with cleaning status label ☐

Production Area Documents

1. Batch manufacturing record ☐
    - Master formula ☐
    - Bill of materials ☐
2. Sanitization record of the area (suspension) ☐

Packaging Area Documents

Batch packaging record

QA In-Process

1. Sampling report (manufacturing side) ☐
2. QA in-process records ☐
    a. Weight variation format (tablets, capsules, suppositories, parenterals) ☐
    b. Inspection start-up checklist with release label ☐
    c. Visual inspection for parenterals ☐
    d. Weighing area checklist signed by QAI ☐

Any Other Documents

1. Process change request ☐
2. Memo (sorting or any other) ☐
3. Waiver document ☐
4. ____________________ ☐

Remarks: ____
____
____

Checked and sent by: ____________ Date: ______________
Audited by: ____________________ Date: ______________
Released by: ___________________ Date: ______________
Quality Unit

Attachment IV
SOP No.: QCS:
Issued on:
Revision No.:
Initiator name:

DOCUMENTS COMPLIANCE DECLARATION

Packaging

Product : ______________________

Code No. : ______________________

Batch No. : ______________________

Consignee/s : ______________________

____

____

This is to confirm that the batch documents sent have been verified against the checklist for the completeness, correctness, clarity, accuracy, and consistency as per the requirement of cGMP and ABC Pharmaceutical Company's standards.

Compliance check by: ____________________

Signature: ___________

Date: ______________

Approved by: ______________________
Packaging manager/designate

Attachment IV
SOP No.: QCS:
Issued on:
Revision No.:
Initiator name:

CHECKLIST FOR THE BATCH DOCUMENT AUDIT

(Packaging Area Documents)

1. Batch packaging record for the presence, correctness of all the entries, and the correctness of manufacturing and expiry dates as per shelf life. ☐
2. Packaging materials requisition (computer copy). ☐
3. Batch yield, final, and at various in-process stage is correct. ☐
4. Rejection verification at various stages of processing and at final stage. ☐
5. Reconciliation of labels, leaflets, and boxes done by the packaging department. ☐
6. Area/equipment line clearance for packaging operation along with machine cleaning status label. ☐
7. Master packaging instructions for different consignee. ☐
8. Output sheet. ☐
9. Repacking addendum (if applicable). ☐
10. Store requisition for the repacking material. ☐
11. First and last label of each roll of the product to be included in the batch file. ☐
12. Batch packaging record for
    a. Blister packing ☐
    b. Suppositories filling and packing ☐
    c. Liquid filling and packing ☐
    d. Powder filling and packing ☐
    e. Bottle packing (tablet/capsule) ☐
    f. Bottle torque (if applicable) ☐
    g. Sachet filling and packing ☐
    h. Hand packing (if applicable) ☐

Any other documents

____

____

____

Reviewed by: ____________________

Applicable & available = [✓]
Applicable not available = [?]
Not applicable = [×]

LEVEL 4
Attachment V
SOP No.:
Issued on:
Revision No.:
Initiator name:

CHECKLIST FOR BATCH DOCUMENTS ALONG WITH SAMPLES (FINISHED PRODUCT)

Product:	____________	**Mfg.:**	____________
Code No.:	____________	**Expiry:**	____________
Batch No.:	____________	**Consignee:**	____________
Stage:	Finished	**Status:**	Partial/Complete

Packaging Area Documents

1. Batch packaging record ☐
2. Batch packaging record
   a. Hand packing ☐
   b. Filling and packing ☐
3. Line clearance format along with machine cleaning status label ☐
4. First signed coded commodities ☐
5. Reconciliation form (labels/cartons) ☐
6. Packaging report and the release of finished product ☐
7. Packaging material requisition ☐
8. Time sheet ☐

QA In-Process

1. Sampling report (manufacturing side) ☐
2. QA in-process record ☐
   a. QA filling chart (cream, ointment, powder filling, suspensions, syrups, drops, blister, or strips (tablet/capsule), sachets) ☐
   b. QA final packaging records ☐
   c. Inspection start-up checklist with release label ☐

Any Other Documents

5. Process change request. ☐
6. Memo (sorting or any other). ☐
7. Repacking request/requisition. ☐
8. ____________ ☐

Remarks: ____
____
____

Checked and sent by: ____________ Date: ______________
Audited by: ____________________ Date: ______________
Released by: ___________________ Date: ______________
Quality Unit

Attachment VI
SOP No.:
Issued on:
Revision No.:
Initiator name:

DOCUMENTS COMPLIANCE DECLARATION

Quality Control

Product : ______________________________

Code No. : ______________________________

Batch No. : ______________________________

CHECKLIST FOR THE BATCH DOCUMENT AUDIT (Quality Control)

1. Test results as reported by the QC analyst are within the specification limits and there is no nonconformance in any of the test. ☐
2. The information related to the batch identity is there and correct. ☐

For All Products

a. Analytical report (bulk product) ☐
b. Microbiological reports ☐

For Parenteral Products

a. Sterility report (for parenteral products) ☐
b. LAL test report ☐
c. Air monitoring record of the filling room ☐
d. Personnel hygiene monitoring report ☐
e. Surface monitoring report ☐
f. Water testing report ☐
g. Gloves print (operator) report ☐
h. Machines swabs report ☐

Any other documents

____

____

____

Compliance check by: ______________________

Signature: ____________

Date: ________________

SOP No.:
Issued on:
Revision No.:
Initiator name:

Annexure VII

QUALITY ASSURANCE DEPARTMENT BATCH DOCUMENTS CHECKLIST (FOR PARENTERAL PRODUCTS)

Product: ____________________

Code No.: ____________ **Mfg.:** ____________

Batch No.: ____________ **Expiry:** ____________

MANUFACTURING, QA (IN-PROCESS), AND QC DOCUMENTATION						
S. No.	**Documents**	**Dept.**	**Yes**	**No**	**N/A**	**Remarks**
1.	Documents Compliance Declaration	Prod.				
2.	Approved Master Formula and Method (MFM) is available, properly filled and signed on each manufacturing step	Prod.				
3.	Batch manufacturing record # 1 and packaging material requisition (for ampoules, vials, rubber stoppers, or flip off seals)	Prod.				
4.	Drug weighing area/equipment clearance checklist with cleaning status label	Prod.				
5.	Raw material weighing record with weighed material printouts with daily weight check for balances printout	Prod.				
6.	Inspection start-up checklist	Prod.				
7.	Area/equipment clearance checklist for each stage/area with release label and equipment cleaning status label	Prod.				
8.	SIP prints for mobile tank	Prod.				
9.	Mobile tank vent filter integrity printout	Prod.				
10.	WFI temperature, conductivity, and pH monitoring printout	Prod.				
11.	WFI printout for each manufacturing stage and final weight printout of bulk solution	Prod.				
12.	Stoppers sterilization chart (if required)	Prod.				
13.	Filtration assemblies/machine parts/gowning sterilization chart	Prod.				
14.	Prefilter integrity test printouts (before and after)	Prod.				

continued

SOP No.:
Issued on:
Revision No.:
Initiator name:

(continued)

MANUFACTURING, QA (IN-PROCESS), AND QC DOCUMENTATION						
S. No.	**Documents**	**Dept.**	**Yes**	**No**	**N/A**	**Remarks**
15.	Sterile filtration integrity test printouts (before and after)	Prod.				
16.	Vials/Ampoules sterilization and depyrogenation chart (tunnel printout)	Prod.				
17.	Unusual interventions record	Prod.				
18.	Ampoules leak test record	Prod.				
19.	Ampoules terminal sterilization printout (if applicable)	Prod.				
20.	Freeze dryer SIP and leak test printouts (if applicable)	Prod.				
21.	Product freeze drying charts (if required)	Prod.				
22.	Freeze vent filters integrity test printouts (if applicable)	Prod.				
23.	Visual inspection report	Prod.				
24.	Disinfectant preparation report	Prod.				
25.	Temperature, RH, and differential pressure trend report	Prod.				
26.	Process deviation reports (if any)	Prod.				
27.	Checklist for bulk batch documents	QA				
28.	Inspection start-up checklist with release label (different stages)	QA				
29.	Bulk sampling reports	QA				
30.	X bar and R control chart	QA				
31.	QA visual inspection report	QA				
32.	Documents compliance declaration	QC				
33.	Analytical report for bulk product	QC				
34.	Physical test reports on bulk solution (if applicable)	QC				
35.	Density/weight determination reports (if applicable)	QC				
36.	Liquid-borne particulate matter test report	QC				
37.	Bulk release form	QC				
38.	Environmental monitoring and personnel hygiene reports	QC				
39.	Micro test reports for bioburden and endotoxin	QC				
	Any other documents					

SOP No.:
Issued on:
Revision No.:
Initiator name:

PACKAGING, QA (IN-PROCESS), AND QC DOCUMENTATION						
S. No.	**Documents**	**Dept.**	**Yes**	**No**	**N/A**	**Remarks**
1.	Master packaging instruction is available and duly signed (different consignees)	Packaging				
2.	Packaging material requisition is attached and appropriately filled and signed (different consignees)	Packaging				
3.	Packaging checklist (Documents Compliance Declaration)	Packaging				
4.	Batch packaging record # 1/addendum BPR	Packaging				
5.	Time sheet for packaging (1st and 2nd shift)	Packaging				
6.	Batch packaging record for hand packing (if applicable)	Packaging				
7.	Printing requisition, line clearance, and reconciliation for labels	Packaging				
8.	Printing requisition, line clearance, and reconciliation for boxes	Packaging				
9.	Reconciliation of leaflets	Packaging				
10.	First signed coded labels, boxes, leaflets	Packaging				
11.	First and last signed label from each used label roll available	Packaging				
12.	Area/equipment/line clearance (packaging and maintenance department) is attached and properly filled for line no.________	Packaging				
13.	Output sheet is available and duly filled and signed	Packaging				
14.	Area/equipment cleaning status labels (different stages)	Packaging				
15.	Packaging reports and release of finished product	Packaging				
16.	Documents compliance declaration	Packaging				
17.	Finished product checklist	QA				
18.	Inspection start-up checklist with release label (each consignee)	QA				
19.	QA in-process final packaging record (each consignee)	QA				
20.	Analytical report for finished product (each consignee)	QC				
	Any other documents					

Audited by: ______________________ Date: ____________

SOP No.:
Issued on:
Revision No.:
Initiator name:

Annexure VIII

QUALITY ASSURANCE DEPARTMENT
BATCH DOCUMENTS CHECKLIST

Product: ______________________________

Code No.: ____________ **Mfg.:** ____________

Batch No.: ____________ **Expiry**: ____________

Documents	Dept.	Yes	No	N/A	Remarks
1. Packaging report and release of finished product (different consignees)	Packaging				
2. QC analytical report of finished product (different consignees)	QC				
3. Finished product checklist	QA				
4. Inspection start-up checklist with release labels (each consignee)	QA				
5. In-process filling chart for tablet and capsules/liquids/ PPS/cream and ointment/sachets/supp.	QA				
6. Final packaging record (for different consignees)	QA				
7. Documents compliance declaration	Packaging				
8. Checklist for batch documents audit	Packaging				
9. Batch packaging record No. 1/addendum BPR	Packaging				
10. Time sheets (1st shift and 2nd shift) for different lines	Packaging				
11. Batch packaging record for blisters/liquids/ ointments and creams/supp./PPS	Packaging				
12. Batch packaging record for bottle torque	Packaging				
13. Batch packaging record for hand packing	Packaging				
14. Reconciliation form for labels (all consignees)	Packaging				
15. First and last signed label from each used label roll available	Packaging				
16. Reconciliation form for boxes (all consignees)	Packaging				
17. Reconciliation form for aluminum/PVC (printed/ coded) (all consignees)	Packaging				

SOP No.:
Issued on:
Revision No.:
Initiator name:

(continued)

Documents	Dept.	Yes	No	N/A	Remarks
18. Reconciliation form for leaflets	Packaging				
19. Reconciliation form for filled bottles/blisters/tubes/supp/inhalers/sachets	Packaging				
20. Area/equipment line clearance with cleaning status labels	Packaging				
21. Line verification checklist signed by Eng. + Pack. department for line no.—	Packaging				
22. Output sheets (1st shift and 2nd shift)	Packaging				
23. Master packaging instructions (different consignees)	Packaging				
24. Packaging materials requisition (different consignees)	Packaging				
25. QC documents compliance declaration	QC				
26. Analytical report for bulk product (core in case of tablet)	QC				
27. Analytical report for bulk product (coating)	QC				
28. Bulk release form (core in case of tablet)	QC				
29. Bulk release form (coating)	QC				
30. Inspection of physical attributes (core)	QC				
31. Inspection of physical attributes (coating)	QC				
32. Bulk product checklist (core + coating)	QA				
33. QA inspection start-up checklist with release label (bulk) (core + coating)	QA				
34. QA in-process X & R chart	QA				
35. Filtration report for liquids	Production				
36. QA sampling report (core + coating)	QA				
37. QA check master print outs	QA				
38. Production documents compliance declaration (core + coating)	Production				
39. Checklist for batch documents audit (core + coating)	Production				
40. Area equipment clearance with cleaning status labels	Production				
41. Drug weighing area/equipment clearance with cleaning status labels	Weighing area				

continued

SOP No.:
Issued on:
Revision No.:
Initiator name:

(continued)

Documents	Dept.	Yes	No	N/A	Remarks
42. Master formula and method (core + coating)	Production				
43. Batch manufacturing record for compression/ encapsulation	Production				
44. Batch manufacturing record for compression (graphical presentation)	Production				
45. Weighing area/equipment checklists with printouts of weighed materials/dispensed liquids from silos with cleaning status labels (core + coating)	Weighing area				
46. Weighing area documents compliance declaration (core + coating)	Weighing area				
47. Checklist for batch documents audit (core + coating)	Weighing area				
48. Batch manufacturing record No. 1 (core + coating)	Weighing area				
49. Disintegration time printout	QA				
50. Temperature and RH and ΔP printout	Production				
51. Time sheet	Production				
52. Mfg. yield form	Production				
Any other documents					

Any batch-related deviation

____

____

____

____

____

Close-out

____

____

____

____

____

Audited by: ____________________

Date: ______________

QAS-001

YOUR COMPANY STANDARD OPERATING PROCEDURE

SOP No. QAS-001.00 Effective date: mm/dd/yyyy
Approved by:

TITLE: **Validation and Certification Guideline**

AUTHOR: ______________________________

Name/Title/Department

____

Signature/Date

CHECKED BY: ______________________________

Name/Title/Department

____

Signature/Date

APPROVED BY: ______________________________

Name/Title/Department

____

Signature/Date

REVISIONS:

No.	Section	Pages	Initials/Date

SOP No. QAS-001.00 Effective date: mm/dd/yyyy
Approved by:

20.1 PURPOSE

To provide a written procedure to be used as a guideline for type of validations, preparation of master validation plan, preparation of validation protocol (installation qualification (IQ), operational qualification (OQ), performance qualification (PQ) and process), and so on.

20.2 RESPONSIBILITIES

1. The QA manager will prepare and update the validation master plan and the validation protocol, report write-up, and maintain the validation documents.
2. The validation team described in the validation master plan will be responsible for performing the validation as per the approved protocol.
3. All departmental managers are responsible for system compliance.
4. The manufacturer or supplier of machines and instrument will be responsible for providing the complete package of IQ and OQ for new equipment related to production, packaging, utilities, and QC.
5. The validation team will be responsible for performing PQ.

20.3 BACKGROUND

Validation is a tool which gives assurance that the manufacturing equipment and process, when operated within a given parameter, will produce the required results consistently. Distinct ongoing benefits can be derived from an effective validation program in the form of increased reliability and productivity, which can result in reduced testing and help avoid nonconformance cost.

20.4 VALIDATION TEAM MEMBERS

A validation team will be formed, one member each from QA, production or packaging, maintenance, QC, and R&D (as per requirement) and members from other concerned departments designated by each departmental head.

20.5 PROCEDURE

20.5.1 Definitions

1. *Master plan:* A comprehensive document describing all the applicable validation requirements for the facility, and providing a plan for meeting those requirements.
2. *Validation:* A systematic approach to gathering and analyzing sufficient data that will give reasonable assurance, based upon scientific judgment, that a manufacturing process, when operating within the specified parameters, will consistently produce product within predetermined specifications.
3. *Qualification:* Verification of the equipment/facility/heating ventilation and air-conditioning (HVAC), and so on, including appropriate supporting systems, can be carried out according to design and/or operating requirements.
4. *Certification:* A review and approval procedure carried out as a final step in the validation program.
5. *Types of validation*
   a. *Prospective:* Analysis of process/equipment prior to use.
   b. *Retrospective:* Validation of a process for a product already in distribution, based on accumulated production, testing and control data.

SOP No. QAS-001.00 Effective date: mm/dd/yyyy
Approved by:

c. *Concurrent:* A combination of retrospective and prospective validation performed against an approved protocol but the product is released on a lot by lot basis. Used on existing product that has not been previously validated or insufficiently validated.
d. *Installation qualification:* When new equipment is installed it should be confirmed that it operates in accordance with design specification or "equipment installed vs design specification." It includes checking the electrical and mechanical input and output flow, directions of valves, operational speed, calibration of control features, and operation of automatic control devices.
e. *Operational qualification:* The purpose of the study is to verify that the equipment is capable of processing at established parameters which will confirm respective specification requirements.
f. *Performance qualification:* To ensure that the system will operate and produce expected product quality results.
g. *Validation master plan preparation:* The purpose of the plan is to
   i. Define all the plant activities, for example, facility, equipment, processes, HVAC system, water system, utilities, cGMP requirements, and products requiring validation.
   ii. Refer all the plans, procedures, and time table intended for accomplishing the validation.

The QA manager shall prepare the validation master plan. The following validation requirements must be considered during the preparation of the validation master plan.

- Definitions of the term "validation"
- What is validation master plan
- Validation team member
- Validation team responsibility
- Concept of qualification/validation
- Facility description
- Equipment description
- HVAC description
- Utilities description
- Validation program overview
  - Validation project management
  - Validation responsibilities
  - Design and validability review
  - Enhanced turn-over package
  - Installation qualification protocols
  - OQ protocols
  - PQ protocols
  - Process validation (PV) protocols
  - Change control initiation
  - Cycle development
  - Validation final reports
  - Validation package
  - Certificate for use in manufacturing
  - Required protocols and procedures

SOP No. QAS-001.00 Effective date: mm/dd/yyyy
Approved by:

- Calibration program summary
- Preventive maintenance program summary
- Key SOPs
- Validation of building
  - Test functions and acceptance criteria
  - Civil work
  - Drainage system
- Validation of utility systems
  - Test functions and acceptance criteria
  - Plant steam (if applicable)
  - Pure steam (if applicable)
  - Water for injection (if applicable)
  - Compressed air (if applicable)
  - Nitrogen (N_2)/carbon dioxide (CO_2) (if applicable)
  - HVAC
  - Emergency power
- Process description
  - Process flow, variables, and responses
- Process descriptions
- Qualification of process equipment
  - Test functions and acceptance criteria
- Validation of support processes, for example:
  - Washing of components
  - Sterilization of components
  - Depyrogenation of components
  - Aseptic filling validation (media fill studies)
  - Cross contamination control
  - Computerized pharmaceutical system
- Test functions and acceptance criteria
- QA/QC laboratory validation
  - Laboratory equipment qualification
  - Computer-related systems in QA/QC
- cGMP procedures and programs
  - Engineering change control
  - Calibration
  - Preventive maintenance program
  - Standard operating procedure (SOPs)
  - Facility cleaning and sanitization
  - Environmental monitoring program
  - High efficiency particulate air (HEPA) filter integrity testing (if applicable)
  - Filter integrity testing (if applicable)
  - Label control program
  - cGMP training
  - Equipment log book, status tags, and room clearance
  - Validation files
  - Risk analysis
  - Validation schedule
  - Validation status
- Drawings/layouts

SOP No. QAS-001.00 Effective date: mm/dd/yyyy
Approved by:

The departmental managers are responsible for checking and reviewing the plan. Directors of R&D, QC, maintenance, and production departments are responsible for the plan approvals. The QA director will sign the plan to give authorization.

20.5.2 Protocols Development

The QA manager or designate will be responsible for the development of a protocol with the consent of validation team members. Selected tests are carried out according to the designed protocol to accumulate data that are subjected to evaluation.

New equipment to be identified in the plan and IQ, OQ, and PQ shall be carried out. Existing equipments revalidation will be carried out as per SOP QAS-000.

20.5.2.1 Installation Qualification

Validation establishes confidence in the process/equipment and ancillary system that they consistently operate within established limits and tolerance and that they are installed properly.

This phase of validation will include the following or as appropriate:

i. Document description
ii. System description
iii. General characteristics
iv. Installation requirements
v. Documentation

For the detailed procedure refer to the corresponding SOP.

20.5.2.2 Operational Qualification

The following elements of the process are to be analyzed at the initiation of the OQ program or as appropriate:

i. Document description
ii. Operational checks
iii. Draft SOPs
iv. Identification of the personnel for training
v. Software/hardware structural testing

For the detailed procedure refer to the corresponding SOP.

20.5.2.3 Performance Qualification

The following elements of the process are to be analyzed at the initiation of the PQ program or as appropriate:

i. Document description
ii. Approved SOPs
iii. Training record for identified personnel
iv. System recovery SOPs
v. PQ test

For the detailed procedure refer to the corresponding SOP.

SOP No. QAS-001.00 Effective date: mm/dd/yyyy
Approved by:

20.5.2.4 Process Validation

The QA manager or designate shall prepare the PV protocol. Existing products (processes) shall be validated by the retrospective approach where possible; for further details, refer to the corresponding SOP.

New processes shall be validated by the prospective approach where possible. Otherwise data should be reviewed after the manufacture of three batches.

20.6 PROTOCOL APPROVAL

Protocols must be approved prior to gathering any data or carrying out any test.

20.7 EXECUTION OF PROTOCOL

Once the protocol is approved, the validation program will proceed as per the procedure outlined in the protocol.

20.8 REVALIDATION

Is most often utilized to validate changes in equipment, packaging, formulation, operating procedure, or the process that could impact the product or product effectiveness or when the occurrence of a problem is identified in the product or process. This is evaluated/revalidated and corrective actions are taken to rectify on the critical system. For revalidation, the same procedure as for validation is followed. For further details of revalidation, refer to the corresponding SOP QAS-000.

20.9 SIGN-OFF REQUIREMENT/APPROVAL

The completed protocols must be reviewed for completeness, accuracy and acceptance of the results and signed off by the QA manager as a minimum.

20.10 VALIDATION DOCUMENTS

The plant QA involved in validation activities shall establish and maintain a central validation document filing system.

These files shall be kept under the control of the QA manager.

These files may include such items as validation protocols, QC testing reports, calibration records, flow diagrams, validation reports, and so on.

20.11 REASON FOR REVISION

First time issued for your company, affiliates, and contract manufacturers.

QAS-002

YOUR COMPANY STANDARD OPERATING PROCEDURE

SOP No. QAS-002.00 Effective date: mm/dd/yyyy
Approved by:

TITLE: **Product Complaints Handling Procedure**

AUTHOR: ______________________________

Name/Title/Department

____

Signature/Date

CHECKED BY: ______________________________

Name/Title/Department

____

Signature/Date

APPROVED BY: ______________________________

Name/Title/Department

____

Signature/Date

REVISIONS:

No.	Section	Pages	Initials/Date

SOP No. QAS-002.00 Effective date: mm/dd/yyyy
Approved by:

Subject: Product Complaints Handling Procedures

21.1 PURPOSE

To provide written instructions for the handling of product complaints. The procedure is designed to ensure that all product complaints are properly and adequately evaluated and are answered in a timely and a professional manner. To ensure that any trend indicating a quality deficiency is noticed at the earliest opportunity.

21.2 RESPONSIBILITY

It is the responsibility of quality affairs director and in his absence the QC director to follow up the procedure.

21.3 PROCEDURE

Note: All complaints shall be replied within 1 week after initial investigation. The investigation must be concluded within 30 days of the receipt of the complaint. In case where the investigation cannot be included within the prescribed time because of some legitimate and justifiable reason/s (including and not limited to: complaint sample not arranged, testing not conclusive, requires repeat testing, etc.) Attachment QAS-040-VI will be raised by the QA director extending the period of investigation further to a maximum of 30 days to conclude the investigation and draft a response for the approval of company's head, if required.

1. All product complaints from each and every source external to the ABC Pharmaceutical Company are to be passed to the QA director. Complaints may be received concerning the quality of products from any source. The QA director will forward the complaint with specific remarks as appropriate for investigation. Complaints may be received from the following possible routes.
   a. Marketing department or sales representatives. For clarity of work, the complaints from marketing and sales shall come on the enclosed form "Customer Quality Complaint Sheet," Attachment QAS-040-I.
   b. Regulatory affairs, via letter, fax, or telephone call from customer or the health authority.
   c. Complaints received from all sources including marketing department are to be assigned a complaint reference number by the QA department (Attachment QAS-040-II: Product Complaint Receipt Form).
   d. A register will also be maintained by the QA to maintain a complaint reference number to all the complaints in sequential order for the year.
2. Prior to any investigation, analysis and reviews for establishing the validity of the complaints, the possible nature of the complaint will be categorized or classified as critical, major, or minor on the basis of the enclosed format Attachment QAS-040-III "Complaint Classification Form."
3. If assistance is required to evaluate the complaint, a meeting will be called by the QA director.
4. After identification of the problem, the investigational work will be carried out.

SOP No. QAS-002.00 Effective date: mm/dd/yyyy
Approved by:

5. The file samples and batch documents will be physically checked and reviewed and if required, chemical or microbiological testing will be performed on reference/complaint samples after approval from the director QA and findings will be recorded on the "Investigation Form" (Attachment QAS-040-IV). Batch document review will also be carried out.
6. Depending on the possible nature of the complaint, finished products, raw materials, or the packaging materials may be reanalyzed by the QC and reported.
7. After all the investigation work is completed, a conclusion will be drawn to ensure the validity of the complaint and to initiate corrective actions appropriate to the nature of the complaint.
8. Quality assurance director is responsible for replying to the complainant.
9. Quality assurance director may suggest calling on legal assistance before framing the response only after consultation with the company's head.
10. If the complaint is of product quality, it may sometimes be compensated by supplying a replacement of the same size and product, together with an appropriate letter of explanation.
11. If the complaint concerned efficacy and side effects, legal advice may be sought before making any form of settlement.
12. All complaints received (Customer Quality Complaint Sheet), review and investigation form along with the corrective action will be maintained in a separate file marked Market Complaints.

21.4 DOCUMENTATION

Attachment I : Customer complaint sheet
Attachment II : Product complaint receipt form
Attachment III : Complaint classification form
Attachment IV : Finished product complaint (investigation form)
Attachment V : Batch document review
Attachment VI : Investigation time extension

21.5 REASON FOR REVISION

First time issued for the ABC Pharmaceutical Company.

21.6 ATTACHMENT FORMS

ABC PHARMACEUTICAL COMPANY

Attachment No.: QAS-002
Revision No.:
Issued on:
Initiator Name:

COMPLAINT INVESTIGATION DURATION EXTENSION FORM

Complaint No.: ______________________
Product: ______________________
Code No: ______________________
Batch No.(s): ______________________
Name of the customer/consignee: ______________________
Country: ______________________
Description of the complaint: ______________________

____

____

____

____

____

____

____

____

Size/intensity: ______________________

____

____

____

____

____

Complaint sample receipt date: ______________________
Complaint receipt date: ______________________
Basis for extension period:
1: ______________________
2: ______________________
3: ______________________
4: ______________________

Time period extended up to: __________ __________ __________
Day **Month** **Year**

____
Signature
QAD

cc: Complaint file

Note: Complaint investigation extension should not exceed 60 working days from the day of the receipt of complaint.

ABC PHARMACEUTICAL COMPANY

Attachment No.: QAS-002
Revision No.:
Issued on:
Initiator Name:

CUSTOMER QUALITY COMPLAINT SHEET
FORM – B

To : Quality assurance director
From : Product manager

Complaint No.: ____________________

Product: ____________ **Form:** ____________ **Concentration:** ________________

Batch No.: ____

Country: ____

Name of the customer: ______________ **Area/city:** __________________

Description of the problem: ____________________________________

____

____

____

____

____

____

____

____

____

____

Size/intensity of the problem: ____________________________________

____

____

____

____

____

____

Collection date of the sample from customer: ___________ / _______ / ___________

Receiving data of the sample by P.M.: ________________ / _______ / ___________

Number of sample sent to quality assurance: ___________ / _______ / ___________

Close out:

____

____

____

____

____

____

Product manager: ________________________ **Director:** ________________

Signature/Date: __________________________ **Signature/Date:** ____________

cc: General manager/materials handling director/production director/QC director/head of product managers/ QC manager

ABC PHARMACEUTICAL COMPANY

Attachment No.: QAS-002
Revision No.:
Issued on:
Initiator Name:

PRODUCT COMPLAINT RECEIPT FORM

Complaint reference number: ____________________

(Assigned by QA Secretary)
Complaint register ref. page: ____________

Complaint: ____________________

____

____

____

____

____

____

Product: ____________________
B. No(s): ____________________
Date received: ____________________
No. of sample received: ____________________

Investigation required

1. **File samples rechecking:** ____________________ ☐
2. **Batch(es) documents review:** ____________________ ☐
3. **Reanalysis (FP, RM, PM):** ____________________ ☐
4. ____________________ ☐
5. ____________________ ☐

____
QA director/QA manager

ABC PHARMACEUTICAL COMPANY

Attachment No.: QAS-002
Revision No.:
Issued on:
Initiator Name:

COMPLAINT CLASSIFICATION FORM

Class I : CRITICAL

Product defects that would result in serious, adverse health consequences or death

1. Active ingredient missing ☐
2. Incorrect dosing of active ingredient that might result in serious adverse health consequences or death ☐
3. Product decomposition that might result in serious, adverse health consequences or death ☐
4. Changes in packaging components that might result in serious, adverse health consequences or death ☐
5. Change in active ingredient release that might result in serious, adverse health consequences or death ☐
6. Unexplainable changes in effectiveness that might result in serious, adverse health consequences or death ☐
7. Misbranding mislabeling or misprints that might result in serious adverse health consequences or death ☐
8. Counterfeit
9. Any other : ______________________________
____
____

Class II: MAJOR

Product defects which will cause no adverse health consequences but will impair efficacy.

ABC PHARMACEUTICAL COMPANY

Attach. No.: QAS-002
Revision No.:
Issued on:
Initiator Name:

Complaint

1. Inactive ingredients ☐
2. Nonpathogenic microorganisms in excess of tolerance limits ☐
3. Excessive amounts of physiologically harmless materials ☐
4. Major and perceptible changes in appearance, odor, taste, physical/technological properties, etc. of the product# ☐
5. Major changes and deviations in packaging components# ☐
6. Misbranding, mislabeling, or misprints of major proportions (e.g., expiry date missing)# ☐
7. Product defect that would impair usability (e.g., broken tablets) ☐
8. Defects in packaging components that would impair usability (e.g., leaks, functional defects) ☐
9. Missing labels ☐
10. Empty packages ☐
11. Underfiling, in terms of volume, weight, or quantity and overweight* ☐
12. Commercial product with sample product identification ☐
13. Labels peeled off, missing inserts* ☐
14. Any other: ______________________________

____

____

Attach. No.: QAS-002
Revision No.:
Issued on:
Initiator Name:

Class III—MINOR

Product defects which will cause no adverse health consequences and will not impair efficacy usefulness.

Complaint

1. Minor deviations from product standards (e.g., in terms of color, odor, taste, shape or occurrence of spots) ☐
2. Standard deviations from packaging component standards (e.g., color, occasional illegible printing of lot number, inconsequential misprints or misprints or mislabeling, such as outdated registration numbers, nomenclatures, etc.)# ☐
3. Packaging defects, such as crooked labels, rubbed-off print, hard to open or loose caps
4. Any other: ______________________

____

____

Remarks: ______________________

____

____

____

____

____

QA director/QA manager

Major in the sense that it could result in misinterpretation or uncertainty on the part of the consumer.

* The extent of the defect is to be considered here. Under certain circumstances, it could be classified as a Class E defect.

FINISHED PRODUCT COMPLAINT INVESTIGATION FORM

Attachment No.: QAS-002
Revision No.:
Issued on:
Initiator Name:

Complaint Ref. No.: ________________
Date:________________

Originator of complaint and title: ____
____
____
____

Method of notification: ____
Reason for complaint: ____
____
____
____
____
____
____

Product name: ____
Code No.: ______________________________
Batch No.(s): ______________________________
Total quantity shipped: ______________________
Quantity involved: __________________________

Attachment No.: QAS-002
Revision No.:
Issued on:
Initiator Name:

Evaluation of complaint

Complaint Sample	File Sample
1. Physical characteristics	
2. Sign of deterioration/problem	
3. Other observation	

Attachment No.: QAS-002
Revision No.:
Issued on:
Initiator Name:

Complaint Sample	Reference/File Sample
4. Packaging/labeling/inserts evaluation	
______________________________ ______________________________ ______________________________ ______________________________ ______________________________ ______________________________ ______________________________	______________________________ ______________________________ ______________________________ ______________________________ ______________________________ ______________________________ ______________________________
5. Assay/chemical analysis	
Initial result/data ______________________________% ______________________________ **Complaint sample** ______________________________ ______________________________ ______________________________ ______________________________ ______________________________	**Retain sample reassay/analysis** ______________________________ ______________________________ ______________________________ ______________________________ ______________________________ ______________________________ ______________________________

QC manager/QC director: ______________________ **Date:** ______________

Attachment No.: QAS-002
Revision No.:
Issued on:
Initiator Name:

BATCH DOCUMENT REVIEW

Quality Control Comments and Suggestion

____

____

____

____

____

Quality assurance evaluation: ______________________________

____

____

____

____

QA manager: ______________________ **Date:** ____________

____

____

____

____

QAD: ______________________ **Date:** ____________

ABC PHARMACEUTICAL COMPANY

Attachment Va
SOP No.: QAS-002
Revision No.:
Issued on:
Initiator sign:

CHECKLIST FOR BATCH DOCUMENTS REVIEW
MARKET COMPLAINT

Cream and Ointment

Product: ________________________________
Code No.: ________________________________
Batch No.: ________________________________

Weighing Area Documents

Check the:

1. Batch manufacturing record for the presence, correctness, and completeness of information regarding the charged quantity and the purchase order no. in the specified columns duly signed. ____________________ ____________________

2. Raw material requisition for the presence, correctness, and completeness of the delivered quantity, QC number, and the signatures of dispensing, issuing, receiving, personnel, and the QAI ____________________ ____________________

3. Raw material weighing record with the printouts of the weighed raw materials ____________________ ____________________

4. Weighing area clearance complete with the signatures of the weighing area officer and the QAI ____________________ ____________________

Reviewed by: ____________________

Attachment Va
SOP No.: QAS-002
Revision No.:
Issued on:
Initiator sign:

Manufacturing/Production Area Documents

Check the:

1. Inspection start-up check list with the release label (manufacturing side) ____________________
2. Batch manufacturing record for ____________________
   a. *Ointment and cream:* On the batch manufacturing record check that ____________________
      i. All the information regarding the identity of the batch are correct ____________________
      ii. Record is duly signed by the manufacturing supervisor ____________________
      iii. Information related to the machine cleaning is complete with the signatures of the worker ____________________
      iv. Yield calculation is correct ____________________
      v. Temperatures (fat phase, water phase), all the timings of (homogenizer, agitator), vacuum limits are present on the record and are within the range specified in the MFM ____________________
3. Was there any deviation from MFM, was any corrective action taken, was it properly documented, was PCR raised and made a part of the Batch document? ____________________

Remarks: ____________________

4. Batch yield final and at various in-process stage is correct ____________________
5. Rejection verification at various stages of processing ____________________

Reviewed by: ____________________

Attachment Va
SOP No.: QAS-002
Revision No.:
Issued on:
Initiator sign:

Packaging Area Documents

Check the:

1. Batch packaging record for the correctness, presence of all the entries, and correctness of the manufacturing and expiry dates as per shelf life ____________________
2. Batch packaging record for ____________________
   a. Ointment and cream ____________________
   b. Tube filling and packing ____________________
   c. Hand packing line (if applicable) ____________________
3. Packaging report and release of finished product (for different consignee) ____________________
4. Packaging materials requisition ____________________
5. Batch yield final and at various in-process stage is correct ____________________
6. Rejection verification at various stages of processing and at final stage ____________________
7. Reconciliation of labels and boxes done by the packaging department ____________________
8. Area/equipment line clearance for packaging operation ____________________
9. MPI for different consignee ____________________
10. Output sheet ____________________
11. Repacking addendum ____________________
12. Store requisition for the repacking material ____________________

Remarks: ____________________

Reviewed by: ____________________

Attachment Va
SOP No.: QAS-002
Revision No.:
Issued on:
Initiator sign:

QA In-Process Documents

Check the:

1. Sampling report (manufacturing side) ____________________
2. Sampling report (packaging side) ____________________
3. QA in-process record of inspection for the filling and packaging of the cream and ointment ____________________
4. Was there any sorting of product ____________________
5. For each consignee, QA final packaging record ____________________
6. Inspection start-up check list for the packaging with release label (packaging side) ____________________
7. First signed coded commodities (aluminum foil, leaflet, box, labels, shipping marks, etc.) ____________________

Reasons: ____________________

Results: ____________________

* In all in-process documents check the presence of hourly checks and confirm that all physical parameters monitored during in-process are within specification limits.

Reviewed by: ____________________

Attachment Va
SOP No.: QAS-002
Revision No.:
Issued on:
Initiator sign:

Quality Control Documents

Check the:

a. Test results as reported by the QC analyst are within specification limits and there is no nonconformance in any of the tests ______________________

b. The information related to batch identity is there and correct ______________________

   1. Analytical report (bulk product) ______________________
   2. Microbiological reports ______________________
   3. Finished product analytical report (For different consignee) ______________________
   4. Any other document related to batch (memo, etc.) ______________________

Details, if any ______________________

Reviewed by: ______________________

ABC PHARMACEUTICAL COMPANY

Attachment Vb
SOP No.: QAS-002
Revision No.:
Issued on:
Initiator sign:

CHECKLIST FOR BATCH DOCUMENTS REVIEW
MARKET COMPLAINT

Powder-Filled Ampoules

Product: ______________________________
Code No.: ______________________________
Batch No.: ______________________________

MANUFACTURING
Production

1. Raw material weighing record ______________________________
2. Weighing print outs ______________________________
3. Drug weighing area/equipment clearance check list ______________________________
4. Temperature and pressure differential records ______________________________
5. Master formula and method ______________________________
6. Manufacturing record No. 1 ______________________________
7. Sterilization record for machine parts, filters, and gowns ______________________________
8. Vials washing and sterilization records ______________________________
9. Area/equipment clearance (manufacturing) ______________________________
10. Area/equipment clearance (visual inspection) ______________________________
11. Visual inspection request from production ______________________________

Reviewed by: ____________________

Attachment Vb
SOP No.: QAS-002
Revision No.:
Issued on:
Initiator sign:

QA In-Process

Check the:

1. Results for visual inspection performed by QAI ______________________
   a. QA in-process filling chart ______________________
   b. Inspection start-up check list ______________________
   c. Release labels ______________________
2. Sampling report ______________________
3. Check list for batch documents (bulk product) ______________________
4. QA final packaging record (for different consignee) ______________________
5. Inspection start-up check list (for different consignee) ______________________
6. Release labels for different consignee ______________________
7. Check list for batch documents (finished product) ______________________

Reviewed by: ______________________

Attachment Vb
SOP No.: QAS-002
Revision No.:
Issued on:
Initiator sign:

FINISHING

Packaging

Check the:

1. Batch packaging record No. 1 ____________________
2. Master packaging instructions ____________________
3. Packaging material requisition ____________________
4. Labels/cartons reconciliation form ____________________
5. Area/equipment clearance (finishing) ____________________
6. Batch packaging record ____________________
7. Time sheet ____________________
8. As in liquid vials ____________________

Reviewed by: ____________________

Attachment Vb
SOP No.: QAS-002
Revision No.:
Issued on:
Initiator sign:

Quality Control

1. Analytical report finished/certificate of analysis ______________________
   a. Sterility report ______________________
   b. LAL test report ______________________
2. Air monitoring record of the filling room ______________________
3. Personnel hygiene monitoring ______________________
4. Surface monitoring ______________________
5. Analytical report (bulk) ______________________
6. Water testing reports ______________________

Others

a. Batch No. interchange request form ______________________

Remarks: ______________________

Reviewed by: ______________________

ABC PHARMACEUTICAL COMPANY

Attachment Vc
SOP No.: QAS-002
Revision No.:
Issued on:
Initiator sign:

CHECKLIST FOR BATCH DOCUMENTS REVIEW
MARKET COMPLAINT

Syrups
Drops
Suspension

Product: ____________________
Code No.: ____________________
Batch No.: ____________________

Weighing Area Documents

Check the:

1. Batch manufacturing for the presence, correctness, and completeness of information related to charged quantity and the purchase order number in the specified columns duly signed ____________________ ____________________
2. Raw material requisition for the presence, correctness, and completeness of delivered quantity and QC number on the record duly signed by the dispensing worker, issuing and receiving personnel and QAI ____________________ ____________________ ____________________
3. Raw material weighing record with the printouts of the weighed raw materials ____________________ ____________________
4. Weighing area clearance complete with the signatures of weighing area officer and the QAI ____________________ ____________________

Reviewed by: ____________________

Attachment Vc
SOP No.: QAS-002
Revision No.:
Issued on:
Initiator sign:

Production Area Documents

Check the:

1. Inspection start-up check list with the release label (manufacturing side) ____________
2. Batch manufacturing record for ____________
   a. *Syrups/Drops/Suspension:* On the batch manufacturing record for syrup/drops/suspension check that ____________
      i. All the information regarding the identity of batch is correct ____________
      ii. Record is duly signed by the manufacturing supervisor ____________
      iii. Information related to vessel cleaning is complete with the signatures of the worker ____________
      iv. Yield calculation is correct ____________
      v. Temperatures of all the various manufacturing stages, all the timings of (homogenizer, agitator), vacuum limits are present on the record and are within the range specified in the MFM ____________
3. Was there any deviation from master manufacturing formula/was any corrective action taken/was it properly documented/was corrective action report raised and made a part of the batch document? ____________

Remarks: ____________

4. Batch yield final and at various in-process stage is correct ____________
5. Rejection verification at various stages of processing and at final stage ____________

Reviewed by: ____________

Attachment Vc
SOP No.: QAS-002
Revision No.:
Issued on:
Initiator sign:

Batch Packaging Records

Check the:

1. Batch packaging record No. 1 for correctness, presence of all the entries, and correctness of manufacturing and expiry date as per shelf life ______________________
2. Batch packaging record for ______________________
   a. Liquid filling and packing ______________________
   b. Hand packing line (if applicable) ______________________
   c. Bottle torque ______________________
3. Packaging report and release of finished product (for different consignees) ______________________
4. Packaging materials requisition ______________________
5. Batch yield final and at various in-process stage is correct ______________________
6. Rejection verification at various stages of processing and at final stage ______________________
7. Reconciliation of labels and boxes done by the packaging department ______________________
8. Area/equipment line clearance for packaging operation ______________________
9. Master packaging instruction for different consignee ______________________
10. Output sheet ______________________
11. Repacking addendum ______________________
12. Store requisition for the repacking material ______________________

Remarks: ______________________

Reviewed by: ______________________

Attachment Vc
SOP No.: QAS-002
Revision No.:
Issued on:
Initiator sign:

QA In-Process Documents

Check the:

1. Sampling report (manufacturing side) ____________________
2. Sampling report (packaging side) ____________________
3. QA in-process record of inspection for filling and packaging of liquids (suspensions, syrups, drops, etc.) ____________________
4. Was there any sorting of product? ____________________
5. For each consignee, QA final packaging record ____________________
6. Inspection start-up check list for packaging with release label (packaging area QAI) ____________________
7. First signed coded commodities (aluminum foil, leaflet, box, labels, shipping marks, etc.) ____________________

Reasons: ____________________

Results: ____________________

* In all in-process documents check the presence of hourly checks and confirm that all the physical parameters are within the specification limits.

Reviewed by: ____________________

Attachment Vc
SOP No.: QAS-002
Revision No.:
Issued on:
Initiator sign:

Quality Control Documents

Check the:

a. Test results as reported by QC analyst are within specification limits and there is no nonconformance in any of the tests ______________________

b. The information related to batch identity is there and correct ______________________

1. Analytical report (bulk product) ______________________
2. Microbiological reports ______________________
3. Finished product analytical report (for different consignee) ______________________
4. Packaging report and release of finished product ______________________
5. Any other document related to batch (memo etc.) ______________________

Details if any ______________________

Reviewed by: ______________________

ABC PHARMACEUTICAL COMPANY

Attachment Vd
SOP No.: QAS-002
Revision No.:
Issued on:
Initiator sign:

CHECKLIST FOR BATCH DOCUMENTS REVIEW
MARKET COMPLAINT

Liquid Vials

Product: ____________________
Code No.: ____________________
Batch No.: ____________________

MANUFACTURING

Production

1. Raw material weighing record ____________________
2. Temperature and pressure differential records ____________________
3. Master formula and method ____________________
4. Manufacturing record No. 1 ____________________
5. Sterilization record for machine parts, filters, and gowns ____________________
6. Vials washing and sterilization records ____________________
7. Area/equipment clearance (manufacturing) ____________________
8. Filter integrity test report ____________________
9. Area/equipment clearance (visual inspection) ____________________
10. Visual inspection request from production ____________________

Reviewed by: ____________________

Attachment Vd
SOP No.: QAS-002
Revision No.:
Issued on:
Initiator sign:

QA In-process

Check the:

1. Results for visual inspection performed by QAI ____________________
2. QA in-process filling chart with inspection start-up check list and release labels ____________________
3. Particle counts records available in logbook ____________________
4. QA final packaging record with inspection start-up check list and release labels (for different consignee) ____________________

Reviewed by: ____________________

FINISHING

Packaging

1. Batch packaging record ____________________
2. Master packaging instructions ____________________
3. Packaging material requisition ____________________
4. Labels/cartons reconciliation form ____________________
5. Area/equipment clearance (finishing) ____________________
6. Batch packaging record ____________________
7. Time sheet ____________________
8. Packaging report and release of finished product (for different consignee) ____________________

Reviewed by: ____________________

Attachment Vd
SOP No.: QAS-002
Revision No.:
Issued on:
Initiator sign:

Quality Control

Check the:

1. Air sampling record of the filling room ______________________
2. Gloves print (operator) ______________________
3. Machine swabs ______________________
4. Water testing reports ______________________
5. Analytical report ______________________

Remarks: ______________________

Reviewed by: ______________________

ABC PHARMACEUTICAL COMPANY

Attachment Ve
SOP No.: QAS-002
Revision No.:
Issued on:
Initiator sign:

CHECKLIST FOR BATCH DOCUMENTS REVIEW
MARKET COMPLAINT

Lyophilized products

Product: ______________________________
Code No.: ______________________________
Batch No.: ______________________________

MANUFACTURING

Production

Check the:

1. Raw material weighing record ______________________________
2. Temperature and pressure differential records ______________________________
3. Master formula and method ______________________________
4. Manufacturing record ______________________________
5. Sterilization cycle chart ______________________________
6. Vials washing and sterilization records ______________________________
7. Lyophilization cycle chart ______________________________
8. Area/equipment clearance (manufacturing) ______________________________
9. Filter integrity test report ______________________________
10. Area/equipment clearance (visual inspection) ______________________________
11. Visual inspection request from production ______________________________

Reviewed by: ____________________

Attachment Ve
SOP No.: QAS-002
Revision No.:
Issued on:
Initiator sign:

QA In-Process

Check the:

1. Results for visual inspection performed by QAI ______________________
2. QA in-process filling chart with inspection start-up check list and release labels. ______________________
3. Particle counts records available in logbook ______________________
4. QA final packaging record with inspection start-up check list and release labels (for different consignee) ______________________

Reviewed by: ______________________

FINISHING

Packaging

1. Batch packaging record ______________________
2. Master packaging instructions ______________________
3. Packaging material requisition ______________________
4. Labels/cartons reconciliation form ______________________
5. Area/equipment clearance (finishing) ______________________
6. Batch packaging record ______________________
7. Time sheet ______________________
8. As in liquid vials ______________________

Reviewed by: ______________________

Attachment Ve
SOP No.: QAS-002
Revision No.:
Issued on:
Initiator sign:

Quality Control

Check the:

1. Analytical report ____________________
2. Air monitoring record of the filling room ____________________
3. Gloves print (operator) ____________________
4. Machines swabs ____________________
5. Water testing reports ____________________

Remarks: ____________________

Reviewed by: ____________________

ABC PHARMACEUTICAL COMPANY

Attachment Vf
SOP No.: QAS-002
Revision No.:
Issued on:
Initiator sign:

CHECKLIST FOR BATCH DOCUMENTS REVIEW
MARKET COMPLAINT

Powder-Filled Vials

Product: ______________________________
Code No.: ______________________________
Batch No.: ______________________________

MANUFACTURING

Production

Check the:

1. Raw material weighing record ______________________________
2. Temperature and pressure differential records ______________________________
3. Master formula and method (MFM) ______________________________
4. Manufacturing record ______________________________
5. Sterilization record for machine parts, filters, and gowns ______________________________
6. Vials washing and sterilization records ______________________________
7. Seals/stoppers sterilization records ______________________________
8. Cramper sterilization record ______________________________
9. Area/equipment clearance (manufacturing) ______________________________
10. Area/equipment clearance (visual inspection) ______________________________
11. Visual inspection request from production ______________________________

Reviewed by: ____________________

Attachment Vf
SOP No.: QAS-002
Revision No.:
Issued on:
Initiator sign:

QA In-Process

Check the:

1. Results for visual inspection performed by QAI ____________________
2. QA in-process filling chart with inspection start-up check list and release labels ____________________
3. Particle counts records available in logbook ____________________
4. Sampling report (bulk) ____________________
5. Check list of documents (bulk) ____________________
6. QA final packaging record with inspection start-up check list and release labels (for different consignee) ____________________

Reviewed by: ____________________

Attachment Vf
SOP No.: QAS-002
Revision No.:
Issued on:
Initiator sign:

Finishing

Packaging

1. Batch packaging record ____________________
2. Master packaging instructions ____________________
3. Packaging material requisition ____________________
4. Labels/cartons reconciliation form ____________________
5. Area/equipment clearance (finishing) ____________________
6. Batch packaging record ____________________
7. Time sheet ____________________
8. As in liquid vials ____________________

Reviewed by: ____________________

Remarks: ____________________

Attachment Vf
SOP No.: QAS-002
Revision No.:
Issued on:
Initiator sign:

Quality Control

Check the:

1. Analytical report (finished) ____________________
2. Air monitoring record of the filling room ____________________
3. Gloves print (operator) ____________________
4. Machine swabs ____________________
5. Water testing reports ____________________
6. Analytical report (bulk) ____________________

Reviewed by: ____________________

ABC PHARMACEUTICAL COMPANY

Attachment Vg
SOP No.: QAS-002
Revision No.:
Issued on:
Initiator sign:

CHECKLIST FOR BATCH DOCUMENTS REVIEW
MARKET COMPLAINT

Suppositories

Product: ______________________________
Code No.: ______________________________
Batch No.: ______________________________

Weighing Area Records

Check the:

1. Batch manufacturing record for the presence, correctness, and completeness of information regarding the charged quantity and the purchase order number in the specified column duly signed ______________________________ ______________________________
2. Raw material requisition for the presence, correctness, and completeness of the delivered quantity and QC number on the record duly signed by the dispensing worker, issuing & dispensing personnel and the QAI ______________________________ ______________________________
3. Raw material weighing record with the printouts of the weighed raw materials ______________________________ ______________________________
4. Weighing area clearance complete with the signatures of the weighing area officer and the QAI ______________________________ ______________________________

Reviewed by: ____________________

Attachment Vg
SOP No.: QAS-002
Revision No.:
Issued on:
Initiator sign:

Production Area Documents

Check the:

1. Inspection start-up check list with the release label (manufacturing side) ______
2. Batch manufacturing record for ______
   a. *Suppositories:* On the batch manufacturing record for suppositories check that ______
      i. All the information regarding the identity of batch is correct ______
      ii. Record is duly signed by the manufacturing supervisor ______
      iii. Information related to machine cleaning is complete with the signatures of the worker ______
      iv. Yield calculation is correct ______
      v. The information and data related to the temperature, mixing time, and mixer speed are within the range specified in the MFM ______
3. Was there any deviation from manufacturing formula and method/master packaging instructions? Was any corrective action taken, was it properly documented, was corrective action report raised and made a part of the batch document? ______
4. Batch yield final and at various in-process stage is correct ______
5. Rejection verification at various stages of processing and at final stage ______

Reviewed by: ______

Remarks: ______

Attachment Vg
SOP No.: QAS-002
Revision No.:
Issued on:
Initiator sign:

Packaging Area Documents

Check the:

1. Batch packaging record for the presence of all the entries and correctness of the manufacturing and expiry date as per shelf-life ______
2. Batch packaging record for ______
   a. Hand packing line (if applicable) ______
   b. Suppositories filling and packing ______
3. Packaging report and release of finished product (for different consignee) ______
4. Packaging materials requisition ______
5. Batch yield final and at various in-process stage is correct ______
6. Rejection verification at various stages of processing and at final stage ______
7. Reconciliation of labels and boxes done by the packaging department ______
8. Area/equipment line clearance for packaging operation ______
9. MPI for different consignee ______
10. Output sheet ______
11. Repacking addendum ______
12. Store requisition for the repacking material ______

Remarks: ______

Reviewed by: ______

Attachment Vg
SOP No.: QAS-002
Revision No.:
Issued on:
Initiator sign:

QA In-Process Documents

Check:

1. Sampling report (manufacturing side) ____________________
2. Sampling report (packaging side) ____________________
3. QA in-process record of inspection for suppositories (weight variation, average wt., physical attributes leak test, etc.). ____________________
4. QA in-process record of inspection for packaging of the suppositories ____________________
5. Was there any sorting of product? ____________________
6. QA final packaging record ____________________
7. Inspection start-up check list for packaging with the release label (packaging side) ____________________
8. First signed coded commodities (aluminum foil, leaflet, box, labels, shipping marks, etc.) ____________________

Reasons: ____________________

Results: ____________________

* In all in process documents check the presence of hourly checks and confirm that all the physical parameters are within the specification limits.

Reviewed by: ____________________

Attachment Vg
SOP No.: QAS-002
Revision No.:
Issued on:
Initiator sign:

Quality Control Documents

Check:

a. Test results as reported by the QC analyst are within the specification limits and there is no nonconformance in any of the tests ______________________

b. The information related to the batch identity is available and correct ______________________

1. Analytical report (bulk product) ______________________
2. Microbiological reports ______________________
3. Finished product analytical report ______________________
4. Any other document related to the batch (memo, deviation report, Q-notes etc.) ______________________

Details if any ______________________

Reviewed by: ______________________

ABC PHARMACEUTICAL COMPANY

Attachment Vh
SOP No.: QAS-002
Revision No.:
Issued on:
Initiator sign:

CHECKLIST FOR BATCH DOCUMENTS REVIEW MARKET COMPLAINT

Ready to Fill Syringes

Product: ______________________
Code No.: ______________________
Batch No.: ______________________

MANUFACTURING

Production

Check the:

1. Raw material weighing record ______________________
2. Temperature and pressure differential records ______________________
3. Master formula and method ______________________
4. Manufacturing record ______________________
5. Sterilization record for machine parts, filters, and gowns ______________________
6. Area/equipment clearance (manufacturing) ______________________
7. Filter integrity test report ______________________
8. Printouts for leak test performed in autoclave ______________________
9. Area/equipment clearance (visual inspection) ______________________
10. Visual inspection request from production ______________________

Reviewed by: ______________________

Attachment Vh
SOP No.: QAS-002
Revision No.:
Issued on:
Initiator sign:

QA In-Process

Check the:

1. Results for visual inspection performed by QA inspector ____________________
2. QA in-process filling chart with inspection start up check list and release labels ____________________
3. Particle counts records available in logbook ____________________
4. QA final packaging record with inspection start up check list and release labels (for different consignee) ____________________

Reviewed by: ____________________

Finishing

Packaging

1. Batch packaging record ____________________
2. Master packaging instructions ____________________
3. Packaging material requisition ____________________
4. Labels/cartons reconciliation form ____________________
5. Area/equipment clearance (finishing) ____________________
6. Batch packaging record ____________________
7. Time sheet ____________________
8. Packaging report and release of finished product (for different consignee) ____________________

Reviewed by: ____________________

Attachment Vh
SOP No.: QAS-002
Revision No.:
Issued on:
Initiator sign:

Quality Control

Check the:

1. Air monitoring record of the filling room ____________________
2. Gloves print (operator) ____________________
3. Machines swabs ____________________
4. Water testing reports ____________________
5. Analytical report ____________________

Remarks: ____________________

Reviewed by: ____________________

ABC PHARMACEUTICAL COMPANY

Attachment Vi
SOP No.: QAS-002
Revision No.:
Issued on:
Initiator sign:

CHECKLIST FOR BATCH DOCUMENTS REVIEW
MARKET COMPLAINT

Tablet and Capsule

Product: ______________________________
Code No.: ______________________________
Batch No.: ______________________________

Weighing Area Documents

Check the:

1. Batch manufacturing record for the presence, correctness, and completeness of information regarding the charge quantity and the purchase order number in the specified columns duly signed ______________________________ ______________________________
2. Raw material requisition for the presence, correctness, and completeness of the delivered quantity, QC number and the signatures of the dispensing, issuing, receiving personnel, and the QAI ______________________________ ______________________________
3. Raw material weighing record with the printouts of the weighed raw material ______________________________ ______________________________
4. Weighing area clearance complete with the signatures of the weighing area in charge and the QAI ______________________________ ______________________________

Reviewed by: ____________________

Attachment Vi
SOP No.: QAS-002
Revision No.:
Issued on:
Initiator sign:

Manufacturing/Production Area Documents

Check the:

1. Inspection start-up check list with the release label (manufacturing side) ______________________
2. Batch manufacturing record for ______________________

   a. *Granulation and final mixing:* On the batch manufacturing record for the granulation and final mixing check that ______________________

      i. All the information regarding the identity of the batch is correct ______________________
      ii. Record is duly signed by the manufacturing supervisor ______________________
      iii. Information related to machine cleaning is complete with the signatures of the worker ______________________
      iv. Yield calculation is correct ______________________
      v. Temperatures and timings mentioned are as per MFM and are matching with the Lytzen Oven Chart (in case if the drying is done on the tray dryer) ______________________

   b. *Capsule filling (wt. variation chart)* ______________________

      i. Information regarding the identity of the batch is correct ______________________
      ii. Record is duly signed by the manufacturing supervisor ______________________
      iii. Information related to the machine cleaning is complete with the signatures of the worker ______________________
      iv. Yield calculation is correct ______________________

Attachment Vi
SOP No.: QAS-002
Revision No.:
Issued on:
Initiator sign:

v. All the entries related to temperature, humidity, and others are recorded and within the range ______________________

vi. Weight variation record of the capsule filling generated by the machine operator is present and within the limits ______________________

c. *Batch manufacturing record for compression and wt. variation chart* ______________________

i. Information regarding the identity of the batch is correct ______________________

ii. Record is duly signed by the manufacturing supervisor ______________________

iii. Information related to the machine cleaning is complete with the signatures of the worker ______________________

vi. Information regarding the identity of the batch is correct ______________________

v. Record is duly signed by the manufacturing supervisor ______________________

vi. Information related to the machine cleaning is complete with the signatures of the worker ______________________

vii. Yield calculation is correct ______________________

viii. All the entries related to the temperature, humidity, and other parameters are recorded and within the range ______________________

ix. Weight variation record of the tablet compression generated by the machine operator is present and is within limits ______________________

x. Batch manufacturing record for coating is present and all the information is correct ______________________

Attachment Vi
SOP No.: QAS-002
Revision No.:
Issued on:
Initiator sign:

3. Was there any deviation from MFM, was any corrective action taken, was it properly documented, was PCR raised and was it made a part of the batch documents? ____________________
4. Batch yield final and at various in-process stage is correct ____________________
5. Rejection verification at various stages of processing and at final stage ____________________
6. Reconciliation of labels and boxes done by the packaging department and verified by the area QAI ____________________

Remarks: ____

Reviewed by: ____________________

Attachment Vi
SOP No.: QAS-002
Revision No.:
Issued on:
Initiator sign:

Packaging Area Documents

Check the:

1. Batch packaging record for the presence, correctness of all the entries, and the correctness of manufacturing and expiry dates as per the shelf-life ______________________
2. Batch packaging record for ______________________
   a. Blister packing ______________________
   b. Hand packing line (if applicable) ______________________
   c. Bottle torque (if applicable) that if the tablet is container pack ______________________
3. Packaging report or release of finished product (for different consignee) ______________________
4. Packaging materials requisition ______________________
5. Batch yield final and at various in-process stage is correct ______________________
6. Rejection verification at various stages of processing and at final stage ______________________
7. Reconciliation of labels and boxes done by the packaging department ______________________
8. Area/equipment line clearance for packaging operation ______________________
9. MPI for different consignee ______________________
10. Output sheet ______________________
11. Repacking Addendum ______________________
12. Store requisition for the repacking material ______________________

Reviewed by: ______________________

Attachment Vi
SOP No.: QAS-002
Revision No.:
Issued on:
Initiator sign:

QA In-Process Documents

Check the:

1. Sampling report (manufacturing side) ______________________
2. Sampling report (packaging side) ______________________
3. QA in-process record of inspection for tablet and capsule (compression and/or encapsulation) ______________________
4. For the presence of hourly checks and confirm that all physical parameters monitored during the in-process including weight variation, thickness, hardness, D.T. friability, temperature, RH, and physical attributes of tablet and capsules are within specification limits ______________________
5. QA in-process record of inspection for filling and packaging of the tablet and capsule ______________________
6. Was there any sorting of product? ______________________
7. For each consignee, QA final packaging record ______________________
8. Inspection start up check list for packaging with release label (packaging side) ______________________
9. First signed coded commodities (aluminum foil, leaflet, box, labels, shipping marks etc.) ______________________

Reasons: ______________________

Results: ______________________

* In all in-process documents check for the presence of hourly checks and confirm that all the physical parameters monitored during in-process are within the specification limits.

Reviewed by: ______________________

Attachment Vi
SOP No.: QAS-002
Revision No.:
Issued on:
Initiator sign:

Quality Control Documents

Check the:

1. Test results as reported by QC analyst are within the specification limits and there is no nonconformance in any of the tests ____________________
2. The information related to the batch identity is there and correct ____________________
   a. Analytical report (bulk product) ____________________
   b. Microbiological reports ____________________
   c. Finished product analytical report (for different consignee) ____________________
   d. Any other document related to batch (Memo etc.) ____________________

Details if any ____________________

Reviewed by: ____________________

ABC PHARMACEUTICAL COMPANY

Attachment Vj
SOP No.: QAS-002
Revision No.:
Issued on:
Initiator sign:

CHECKLIST FOR BATCH DOCUMENTS REVIEW
MARKET COMPLAINT

Granules

Product: ______________________________
Code No.: ______________________________
Batch No.: ______________________________

Weighing Area Documents
Check the:

1. Batch manufacturing record for the presence, correctness, and completeness of information regarding the charged quantity and the purchase order number in the specified columns duly signed ______________________________ ______________________________
2. Raw material requisition for the presence, correctness, and completeness of the delivered quantity, QC number and the signatures of the dispensing, issuing, receiving, personnel, and the QAI ______________________________ ______________________________
3. Raw material weighing record with the printouts of the weighed raw material ______________________________ ______________________________
4. Weighing area clearance complete with the signatures of the weighing area officer and the QAI ______________________________ ______________________________

Reviewed by: ____________________

Attachment Vj
SOP No.: QAS-002
Revision No.:
Issued on:
Initiator sign:

Production Area Documents

Check the:

1. Inspection start-up check list with the release label (manufacturing side) ______________________
2. Batch manufacturing record for ______________________
   a. *Granulation and final mixing:* On the batch manufacturing record for the granulation and final mixing check that ______________________
      i. All the information regarding the identity of batch is correct ______________________
      ii. Record is duly signed by the manufacturing supervisor ______________________
      iii. Information related to machine cleaning is complete with the signatures of the worker ______________________
      iv. Yield calculation is correct
      v. Temperatures and timings mentioned are as per MFM and are matching with the Lytzen Oven Chart (in case if the drying is done on the tray dryer) ______________________
   b. *Charts* (*if drying on Lytzen dryer*): Check the presence of the Lytzen dryer charts if the drying of granules was carried out in the Lytzen Oven and the temperature at which the granules were dried was comparable to the limits provided in the MFM ______________________
3. Was there any deviation from MFM, was any corrective action taken, was it properly documented, was PCR raised and made a part of the batch document? ______________________

Remarks: ______________________

4. Batch yield final and at various in-process stage is correct ______________________
5. Rejection verification at various stages of processing and at final stage ______________________

Reviewed by: ______________________

Attachment Vj
SOP No.: QAS-002
Revision No.:
Issued on:
Initiator sign:

Packaging Area Documents

Check the:

1. Batch packaging record for the correctness, presence of all the entries, and correctness of manufacturing and expiry date as per shelf life ______
2. Batch packaging record for ______
   a. Hand packing line ______
   b. Bottle torque ______
   c. Fill wt/powder filling ______
3. Packaging report and release of finished product (for different consignee) ______
4. Packaging material requisition ______
5. Batch yield final and at various in-process stage is correct ______
6. Rejection verification at various stages of processing and at the final stage ______
7. Reconciliation of labels and boxes done by the packaging department ______
8. Area/equipment line clearance for packaging operation ______
9. MPI for different consignee ______
10. Output sheet ______
12. Repacking addendum ______
13. Store requisition for the repacking material ______

Remarks: ______

Reviewed by: ______

Attachment Vj
SOP No.: QAS-002
Revision No.:
Issued on:
Initiator sign:

QA In-Process Documents

Check the:

1. Sampling report (manufacturing side) ____________________
2. Sampling report (packaging side) ____________________
3. QA in-process record of inspection for filling and packaging of PPS ____________________
4. Was there any sorting of product? ____________________
5. For each consignee, QA final packaging record ____________________
6. Inspection start-up check list for packaging with release label (packaging side) ____________________
7. First signed coded commodities (aluminum foil, leaflet, box, labels, shipping marks, etc.) ____________________

Reasons: ____________________

Results: ____________________

* In all in-process documents check the presence of hourly checks and confirm that all the physical parameters monitored during in-process are within the specification limits.

Reviewed by: ____________________

Attachment Vj
SOP No.: QAS-002
Revision No.:
Issued on:
Initiator sign:

Quality Control Documents

Check the:

1. Test results as reported by the QC analyst are within the specification limits and there is no nonconformance in any of the tests ______________________
2. The information related to batch identity is there and correct ______________________
   a. Analytical report (bulk product) ______________________
   b. Microbiological reports ______________________
   c. Finished product analytical report (for different consignee) ______________________
   d. Any other document related to batch (memo, etc.) ______________________

Details, if any ______________________

Reviewed by: ______________________

CANCELLED FORM

CHECK LIST FOR THE BATCH DOCUMENT REVIEW MARKETING COMPLAINT

Cream and Ointment

Product: ______________________________
Code No.: ______________________________
Batch No.: ______________________________

Weighing Area Documents

Check the:

1. Batch manufacturing record for the presence, correctness, and completeness of information regarding the charged quantity and the purchase order number in the specified columns duly signed ______________________________ ______________________________
2. Raw material requisition for the presence, correctness, and completeness of the delivered quantity, QC number, and the signatures of dispensing, issuing, receiving, personnel, and the QAI ______________________________ ______________________________
3. Raw material weighing record with the printouts of the weighed raw materials ______________________________ ______________________________
4. Weighing area clearance complete with the signatures of the weighing area officer and the QAI ______________________________ ______________________________

Reviewed by: ____________________

Manufacturing/Production Area Documents

Check the:

1. Inspection start-up check list with the release label (manufacturing side) ______
2. Batch manufacturing record for ______

   a. *Ointment and cream:* On the batch manufacturing record check that ______

      i. All the information regarding the identity of the batch are correct ______

      ii. Record is duly signed by the manufacturing supervisor ______

      iii. Information related to the machine cleaning is complete with the signatures of the worker ______

      iv. Yield calculation is correct ______

      v. Temperatures (fat phase, water phase), all the timings of (homogenizer, agitator), vacuum limits are present on the record and are within the range specified in the MFM ______

3. Was there any deviation from MFM, was any corrective action taken, was it properly documented, was PRC raised and made a part of the batch document? ______

Remarks: ______

4. Batch yield in the final and at various in-process stage is correct ______
5. Rejection verification at various stages of processing ______

Reviewed by: ______

Packaging Area Documents

Check the:

1. Batch packaging record No. 1 for the correctness, presence of all the entries, and correctness of the manufacturing and expiry dates as per the shelf life ______________________
2. Batch packaging record for ______________________
   a. Ointment and cream ______________________
   b. Tube filling and packing ______________________
   c. Hand packing line (if applicable) ______________________
3. Packaging report and release of finished product (for different consignee) ______________________
4. Packaging materials requisition ______________________
5. Batch yield final and at various in-process stage is correct ______________________
6. Rejection verification at various stages of processing and at final stage ______________________
7. Reconciliation of labels and boxes done by the packaging department ______________________
8. Area/equipment line clearance for packaging operation ______________________
9. Master packaging instructions ______________________
10. Output sheet ______________________
11. Repacking addendum ______________________
12. Store requisition for the repacking material ______________________

Remarks: ____

Reviewed by: ______________________

QA In-Process documents

Check the:

1. Sampling report (manufacturing side) ____________________
2. Sampling report (packaging side) ____________________
3. QA in-process record of inspection for the filling and packaging of the cream and ointment ____________________
4. Was there any sorting of product? ____________________
5. For each consignee, QA final packaging record ____________________
6. Inspection start-up check list for the packaging with release label (packaging side) ____________________
7. First signed coded commodities (aluminum foil, leaflet, box, labels, shipping marks, etc.) ____________________

Reasons: ____________________

Results: ____________________

* Check all in-process documents for the presence of hourly checks and confirm that all physical parameters monitored during in-process are within the specification limits.

Reviewed by: ____________________

Quality Control Documents

Check the:

a. Test results as reported by the QC analyst are within the specification limits and there is no nonconformance in any of the tests ____________________

b. The information related to batch identity is there and correct ____________________

1. Analytical report (bulk product) ____________________

2. Microbiological reports ____________________

3. Finished product analytical report (for different consignee) ____________________

4. Any other document related to batch (memo, etc.) ____________________

Details if any ____________________

Reviewed by: ____________________

QAS-003

YOUR COMPANY STANDARD OPERATING PROCEDURE

SOP No. QAS-003.00 Effective date: mm/dd/yyyy
Approved by:

TITLE: **Sampling of Finished Product for Laboratory Testing**

AUTHOR: ______________________________

Name/Title/Department

____

Signature/Date

CHECKED BY: ______________________________

Name/Title/Department

____

Signature/Date

APPROVED BY: ______________________________

Name/Title/Department

____

Signature/Date

REVISIONS:

No.	Section	Pages	Initials/Date

SOP No. QCS-003.00 Effective date: mm/dd/yyyy
Approved by:

22.1 PURPOSE

To provide written instructions to the QA in-process inspector to draw finished product samples as per written instructions.

22.2 RESPONSIBILITY

It is the responsibility of the QAI to follow the procedure. The QA manager is responsible for the SOP compliance.

22.3 PROCEDURE

1. The QAI will be available in the packaging area. He will collect samples from each running line in the final packaging for both testing and reference purposes.
2. A number of samples will be collected as per the attached annexure.
3. The samples to be collected for each finished product will be divided over the process of packaging and will be collected hourly during audit checks, so that they are representative of the entire packaging process.
4. Injectables: For the sampling of injectables, the process is as follows:
   a. The QAI will collect the samples from the injectables finishing line.
   b. The number of samples will be decided as per the attached annexure.
5. The QAI for in-process control will send the following documents to the QC administration according to the checklist; refer to SOP QAS-000.
6. The samples to be used as reference samples will be kept in alphabetical order in the reference sample room and will be used in case of any complaints from pharmacies or customers for reanalysis by the QC laboratory.
7. They will be stored for the time of the shelf life plus 1 year.

22.4 REASON FOR REVISION

First time issued for the ABC Pharmaceutical Company.

22.5 ATTACHMENT FORM

ABC PHARMACEUTICAL COMPANY

LEVEL 3
SOP No.: QAS-003
Issued on:
Revision No.:
Initiator name:

Annexure I
Sampling Stage and Sample Amount for Different Products

S. No.	Product Name	Sampled Quantity and Type of Container							
		Bulk (g)	Chemical Bulk	Container	Physical		Microbiological		
					Bulk	Finished	Bulk	Finished	Container
1.	Paracetamol 125 mg	–	1 × 100 Supp.	FB	–	2 boxes (40 blisters)	–	1 × 100 Supp. per day	FP
2.	Paracetamol 250 mg	–	1 × 100 Supp.	FB	–	2 boxes (40 blisters)	–	1 box/100 Supp. per day	FP
3.	Paracetamol 500 mg	–	1 × 100 Supp.	FB	–	2 boxes (40 blisters)	–	1 box/100 Supp. per day	FP
4.	Diclofenac 12.5 mg	–	1 × 100 Supp.	FB	–	2 boxes (40 blisters)	–	1 box/100 Supp. per day	FP
5.	Diclofenac 25 mg	–	1 × 100 Supp.	FB	–	2 boxes (40 blisters)	–	1 box/100 Supp. per day	FP
6.	Diclofenac 50 mg	–	1 × 100 Supp.	FB	–	2 boxes (40 blisters)	–	1 box/100 Supp. per day	FP
7.	Paracetamol 120 mg suspension	–	4 × 200 mL	ACB	–	20 bottles	FP	Paracetamol 120 mg suspension	–
8.	Paracetamol 250 mg suspension	–	4 × 200 mL	ACB	–	20 bottles	FP	Paracetamol 250 mg suspension	–
9.	Metronidazole 125 mg/5 mL	–	4 × 200 mL	ACB	–	20 bottles	FP	Metronidazole 125 mg/5 mL	–
10.	Ciprofloxacin 200 mg/5 mL	–	4 × 200 mL	ACB	–	20 bottles	FP	Ciprofloxacin 200 mg/5 mL	–
11.	Fusidic acid 2%	–	100 g	PC	–	20 tubes	Tubes	1 × 50 g	PC
12.	Mikacin sulfate	–	100 g	PC	–	20 tubes	Tubes	1 × 50 g	PC

continued

(continued)

S. No.	Product Name	Sampled Quantity and Type of Container							
		Bulk (g)	Chemical Bulk	Container	Physical		Microbiological		
					Bulk	Finished	Bulk	Finished	Container
13.	Ceuzime 250 mg	–	6 × 100 g	PB	–	30 blister	FP	100–150 g	PB
14.	Cefuzime 500 mg	–	6 × 100 g	PB	–	30 blister	FP	100–150 g	PB
15.	Diclofenac 100 mg C.R.	–	100 g per cont.	PB	–	30 blister	FP	100–150 g	PB
16.	Betamethasone skin	–	100 g	PC	–	20 tubes	Tubes		PC
17.	Amoxycilline/clavulanate 250 mg	–	6 × 100 g	PB	–	30 blister	FP	100–150 g	PB
18.	Amoxycilline/clavulanate 500 mg	–	6 × 100 g	PB	–	30 blister	FP	100–150 g	PB
19.	Penicillin 250 mg	–	6 × 100 g	PB	–	30 blister	FP	100–150 g	PB
20.	Penicillin 500 mg	–	6 × 100 g	PB	–	30 blister	FP	100–150 g	PB
21.	Mix vitamin-M tablets	–	30–60 blister	PB	–	40 blister	FP		PB
22.	Omeprazole 10 mg	–	50 g per cont.	PB	–	20–30 cont. (30 blister)	FP		PB
23.	Omeprazole 40 mg	–	50 g per cont.	PB	–	20–30 cont. (30 blister)	FP		PB
24.	Omeprazole 20 mg	–	50 g per cont.	PB	–	30 blister	FP		PB
25.	Azithromycin 250 mg	–	6 × 100 g	PB	–	30 blister	FP		PB
26.	Simvastatin 15 mg	–	6 × 100 g	PB	–	30 blister	FP		PB
27.	Simvastatin 30 mg	–	6 × 100 g	PB	–	30 blister	FP		PB
28.	Folic acid/iron	–	6 × 100 g	PB	–	30 blister	FP		PB
29.	Diclofenac creamagel	–	100 g	PC	–	20 tubes	FP	–	–
30.	Betamethasone skin cream	–	100 g	PC	–	20 tubes	FP	–	–
31.	Ibuprofin 5%	–	100 g	PC	–	20 tubes per consignee	FP	1 × 100 g	PC
32.	Fusidic acid 2%	–	100 g	PC	–	20 tubes per consignee	FP	1 × 100 g	PC
33.	Silver sulfadiazin 1% cream	–	100 g	PC	–	20 tubes per consignee	FP	1 × 100 g	PC

34.	Pediatric cough syrup	–	2 × 200 mL	ACB	2 × 100 m	20 bottles	FP	–	5 bottles per day	–
35.	Adult cough syrup	–	2 × 200 mL	ACB	2 × 100 m	20 bottles	FP	–	5 bottles per day	–
36.	Salbutamol syrup	–	2 × 200 mL	ACB	2 × 100 m	20 bottles	FP	–	5 bottles per day	–
37.	Flu syrup	–	2 × 200 mL	ACB	2 × 100 m	20 bottles	FP	–	5 bottles per day	–
38.	Pseudoephedrin syrup	–	2 × 200 mL	ACB	2 × 100 m	20 bottles	FP	–	5 bottles per day	–
39.	Pseudoephedrin/ dextromethorphan syrup	–	2 × 200 mL	ACB	2 × 100 m	20 bottles	FP	–	5 bottles per day	–
40.	Cetrizidine 5 mg/5 mL syrup	–	2 × 200 mL	ACB	2 × 100 m	20 bottles	FP	–	5 bottles per day	–
41.	Loratadin 5 mg/5 mL syrup	–	2 × 200 mL	ACB	2 × 100 m	20 bottles	FP	–	5 bottles per day	–
42.	Hyoscine-*N*-butylbromide syrup	–	2 × 200 mL	ACB	2 × 100 m	20 bottles	FP	–	5 bottles per day	–
43.	Oxytetrazoline 0.025% drops	–	2 × 200 mL	ACB	3 × 100 m	20 bottles	FP	–	10 bottles per day	–
44.	Oxytetrazoline 0.05% drops	–	2 × 200 mL	ACB	3 × 100 m	20 bottles	FP	–	10 bottles per day	–
45.	Vitamin B complex tablet	–	10–14 × 100 g	PB	–	30 blisters	FP	–	10 blisters	–
46.	Betamethasone tablet	–	3–4 × 80 g	PB	–	30 blisters	FP	–	10 blisters	–
47.	Salbutamol 2 mg	–	3–4 × 80 g	PB	–	30 blisters	FP	–	10 blisters	–
48.	Salbutamol 4 mg	–	3–4 × 80 g	PB	–	30 blisters	FP	–	10 blisters	–
49.	Cimetidine 800 mg	–	3–6 × 100 g	PB	–	30 blisters	FP	–	10 blisters	–
50.	Cimetidine 200 mg	–	3–6 × 100 g	PB	–	30 blisters	FP	–	10 blisters	–

continued

(continued)

S. No.	Product Name	Sampled Quantity and Type of Container								
					Physical				Microbiological	
		Bulk (g)	Chemical Bulk	Container	Bulk	Finished		Bulk	Finished	Container
51.	Cimetidine 400 mg	–	3–6 × 100 g	PB	–	30 blisters	FP	–	10 blisters	–
52.	Diclofenac 25 mg tablets	–	4–8 × 100 g	PB	–	30 blisters	FP	–	10 blisters	–
53.	Erythromycin 500 mg	–	2–4 × 100 g	PB	–	30 blisters	FP	–	10 blisters	–
54.	Famotidine 20 mg	–	2–4 × 100 g	PB	–	30 blisters	FP	–	10 blisters	–
55.	Famotidine 40 mg	–	2–4 × 100 g	PB	–	30 blisters	FP	–	10 blisters	–
56.	Anti flu	–	7–10 × 100 g	PB	–	30 blisters	FP	–	10 blisters	–
57.	Folic acid 5 mg	–	6–10 × 100 g	PB	–	30 blisters	FP	–	10 blisters	–
58.	Gliclizide	–	3–6 × 80 g	PB	–	30 blisters	FP	–	10 blisters	–
59.	Ibuprofen 600	–	5–10 × 100 g	PB	–	30 blisters	FP	–	10 blisters	–
60.	Ibuprofen 200	–	5–10 × 100 g	PB	–	30 blisters	FP	–	10 blisters	–
61.	Ibuprofen 400	–	10–14 × 100 g	PB	–	30 blisters	FP	–	10 blisters	–
62.	Ranitidine HCl 150 mg	–	3–6 × 100 g	PB	–	30 blisters	FP	–	10 blisters	–
63.	Ranitidine HCl 300 mg	–	3–6 × 100 g	PB	–	30 blisters	FP	–	10 blisters	–
64.	Hyoscine-*N*-butyl bromide	–	3–6 × 80 g	PB	–	30 blisters	FP	–	10 blisters	–
66.	Aspirin	–	6 × 100 g	PB	–	30 blisters	FP	–	10 blisters	–
67.	Diclofenac	–	6 × 100 g	PB	–	30 blisters	FP	–	10 blisters	–

Sampling Stage and Sample Amount for Different Products

S. No.	Product Name	Sampled Quantity and Type of Container							
					Physical		Microbiological		
		Bulk (g)	Chemical Bulk	Container	Bulk	Finished	Bulk	Finished	Container
1.	Paracetamol 500 mg	–	7–10 × 100 g	PB	–	30 blisters	FP	–	10 blisters
2.	Paracetamol compound	–	7–10 × 100 g	PB	–	30 blisters	FP	–	10 blisters
3.	Vitamin B complex tablet	–	10–14 × 100 g	PB	–	30 blisters	FP	–	10 blisters
4.	Betamethasone tablet	–	3–4 × 80 g	PB	–	30 blisters	FP	–	10 blisters
5.	Salbutamol 2 mg	–	3–4 × 80 g	PB	–	30 blisters	FP	–	10 blisters
6.	Salbutamol 4 mg	–	3–4 × 30 g	PB	–	30 blisters	FP	–	10 blisters
7.	Cimetidine 800 mg	–	3–6 × 100 g	PB	–	30 blisters	FP	–	10 blisters
8.	Cimetidine 200 mg	–	3–6 × 100 g	PB	–	30 blisters	FP	–	10 blisters
9.	Cimetidine 400 mg	–	3–6 × 100 g	PB	–	30 blisters	FP	–	10 blisters
10.	Diclofenac 25 mg tablets	–	4–8 × 100 g	PB	–	30 blisters	FP	–	10 blisters
11.	Diclofenac 50 mg tablets	–	4–8 × 100 g	PB	–	30 blisters	FP	–	10 blisters
12.	Erythromycin 500 mg	–	2–4× 100 g	PB	–	30 blisters	FP	–	10 blisters
13.	Famotidine 20 mg	–	2–4 × 100 g	PB	–	30 blisters	FP	–	10 blisters
14.	Famotidine 40 mg	–	2–4 × 100 g	PB	–	30 blisters	FP	–	10 blisters
15.	Anti flu	–	7–10 × 100 g	PB	–	30 blisters	FP	–	10 blisters
16.	Folic acid 5 mg	–	6–10 × 100 g	PB	–	30 blisters	FP	–	10 blisters
17.	Gliclizide	–	3–6 × 80 g	PB	–	30 blisters	FP	–	10 blisters
18.	Ibuprofen 600	–	5–10 × 100 g	PB	–	30 blisters	FP	–	10 blisters
19.	Ibuprofen 200	–	5–10 × 100 g	PB	–	30 blisters	FP	–	10 blisters

continued

(continued)

S. No.	Product Name	Sampled Quantity and Type of Container							
					Physical		Microbiological		
		Bulk (g)	Chemical Bulk	Container	Bulk	Finished	Bulk	Finished	Container
20.	Ibuprofen 400	–	10–14 × 100 g	PB	–	30 blisters	FP	–	10 blisters
21.	Ranitidine HCl 150 mg	–	3–6 × 100 g	PB	–	30 blisters	FP	–	10 blisters
22.	Ranitidine HCl 300 mg	–	3–6 × 100 g	PB	–	30 blisters	FP	–	10 blisters
23.	Hyoscine-*N*-butyl bromide	–	3–6 × 80 g	PB	–	30 blisters	FP	–	10 blisters
24.	Pseudoephedrine	–	3–6 × 80 g	PB	–	30 blisters	FP	–	10 blisters
25.	Vitamin C 100 mg	–	4–8 × 100 g	PB	–	30 blisters	FP	–	10 blisters
26.	Vitamin C 500 mg	–	6–8 × 100 g	PB	–	30 blisters	FP	–	10 blisters
27.	Aspirin	–	6 × 100 g	PB	–	30 blisters	FP	–	10 blisters
28.	Diclofenac	–	6 × 100 g	PB	–	30 blisters	FP	–	10 blisters
29.	Acyclovir	–	6 × 100 g	PB	–	30 blisters	FP	–	10 blisters
30.	Clindamycin	–	6 × 100 g	PB	–	30 blisters	FP	–	10 blisters
31.	Amoxycillin/clavulanate 375 mg	–	6 × 100 g	PB	–	30 blisters	FP	6 × 100 g	10 blisters
32.	Ciprofloxacin 250 mg	–	6 × 100 g	PB	–	30 blisters	FP	6 × 100 g	10 blisters
33.	Ciprofloxacin 500 mg	–	6 × 100 g	PB	–	30 blisters	FP	6 × 100 g	10 blisters
34.	Ciprofloxacin 750 mg	–	6 × 100 g	PB	–	30 blisters	FP	6 × 100 g	10 blisters
35.	Levofloxacin 400 mg	–	6 × 100 g	PB	–	30 blisters	FP	–	10 blisters
36.	Metformin 500 mg film coated	–	6 × 100 g	PB	–	30 blisters	FP	–	10 blisters
37.	Cetrizidine 10 mg film coated	–	6 × 100 g	PB	–	30 blisters	FP	–	10 blisters
38.	Loratidine 10 mg	–	6 × 100 g	PB	–	30 blisters	FP	–	10 blisters
39.	Metronidazole	–	6 × 100 g	PB	–	30 blisters	FP	–	10 blisters

Sampling Stage and Sample Amount for Different Products

S. No.	Product Name	Chemical Bulk (g/mL)	Contai-ner	Micro Bulk (g/mL)	Contai-ner	F. Chemical Bulk after Leak Test Autoclave Lyophilization	Contai-ner	Micro F. Bulk after Leak Test Autoclave Lyophilization	Contai-ner	Physical	Contai-ner	After Visual Inspection for Liquid-Borne Particulate Matter
Injections												
1.	Diclofenac 75 mg/3 mL	50 mL (C-19/2)	ACB	100 mL (C-19/2 and C-31)	SPFB	20 Ampoules	PB	40 Ampoules	PB	Pack size 50s 1 box Pack size 5s 4 boxes Pack size 3s 3 boxes	FP	10 Ampoules
2.	SWFI 10 mL SWFI 5 mL SWFI 4 mL SWFI 2 mL	400 mL (C-19/2) 400 mL (C-19/2) 400 mL (C-19/2) 400 mL (C-19/2)	ACB ACB ACB ACB	200 mL + 100 mL (C-19/2 and C-31) 200 mL + 100 mL (C-19/2 and C-31) 200 mL + 100 mL (C-19/2 and C-31) 200 mL + 100 mL (C-19/2 and C-31)	SPFB SPFB SPFB SPFB	40 Ampoules 80 Ampoules 100 Ampoules 200 Ampoules	PB PB PB PB	40 Ampoules 40 Ampoules 40 Ampoules 40 Ampoules	PB PB PB PB	Pack size 50s 1 box Pack size 50s 1 box Pack size 50s 1 box Pack size 50s 1 box	FP	10 Vials
3.	Vancomycin HCl 0.5 g	50 mL (C-19/2)	ACB	100 mL (C-19/2 and C-31)	SPFB	20 Vials	PB	40 Vials	PB	Pack size 10s 1 box Pack size 1s 20 boxes	FP	10 Vials
4.	Metaclopromide 10 mg/2 mL	50 mL (C-19/2)	ACB	100 mL (C-19/2 and C-31)	SPFB	20 Ampoules	PB	40 Ampoules	PB	Pack size 50s 1 box Pack size 5s 4 boxes Pack size 3s 3 boxes	FP	10 Ampoules

continued

(continued)

S. No.	Product Name	Chemical Bulk (g/mL)	Container	Micro Bulk (g/mL)	Container	F. Chemical Bulk after Leak Test Autoclave Lyophilization	Container	Micro F. Bulk after Leak Test Autoclave Lyophilization	Container	Physical	Container	After Visual Inspection for Liquid-Borne Particulate Matter
5.	Omeprazole 40 mg	50 mL (C-19/2)	ACB	100 mL (C-19/2 and C-31)	SPFB	20 Vials	PB	40 Vials	PB	Pack size 10s 1 box Pack size 1s 20 boxes	FP	10 Vials
6.	Lidocaine HCl Inj. 1% - 2 mL - 3 mL - 3.5 mL - 4 mL	50 mL (C-19/2)	ACB	100 mL (C-19/2 and C-31)	SPFB	20 Ampoules	PB	40 Ampoules	PB	–	–	–
7.	Ranitidine HCl 50 mg/2 mL	50 mL (C-19/2)	ACB	100 mL (C-19/2 and C-31)	SPFB	20 Ampoules	PB	40 Ampoules	PB	Pack size 50s 1 box Pack size 5s 4 boxes Pack size 3s 3 boxes	FP	10 Ampoules
8.	Hyoscine-*N*-butylbromide 20 mg/mL	50 mL (C-19/2)	ACB	100 mL (C-19/2 and C-31)	SPFB	20 Ampoules	PB	40 Ampoules	PB	Pack size 50s 1 box Pack size 5s 4 boxes Pack size 3s 3 boxes	FP	20 Ampoules

9.	Vitamin B-complex	50 mL (C-19/2)	ACB	100 mL (C-19/2 and C-31)	SPFB	20 Ampoules	PB	40 Ampoules	PB	Pack size 50s 1 box Pack size 5s 4 boxes Pack size 3s 3 boxes	FP	20 Ampoules
10.	Erythropoietin 2000 IV	50 mL (C-19/2)	ACB	100 mL (C-19/2 and C-31)	SPFB	20 Vials	PB	40 Vials	PB	Pack size 10s 1 box	FP	20 Vials
11.	Erythropoietin 4000 IV	50 mL (C-19/2)	ACB	100 mL (C-19/2 and C-31)	SPFB	20 Vials	PB	40 Vials	PB	Pack size 10s 1 box	FP	10 Vials
12.	Amoxycillin 0.6 g powder for I.U. injection	–	–	–	–	15 Vials	PB	46 Vials	PB	Pack size 10s 1 box	FP	
13.	Amoxycillin 1.2 g powder for I.U. injection	–	–	–	–	15 Vials	PB	46 Vials	PB	Pack size 10s 1 box	FP	
14.	ECOCAIN 2% 1.100,000	–	–	–	–	40 Dental cartridge	PB	46 Dental cartridge	PB	Pack size 50s 1 box Pack size 10s 1 box	FP	10 Cartridge
15.	Calcitriol	50 mL (C-19/2)	ACB	1) 100 mL (C-19/2) 2) 100 mL (C-31)	SPFB SPFB	20 Ampoules	PB	40 Ampoules	PB	Pack size 5s 20 boxes	FP	10 Ampoules
16.	Juslin R	50 mL (C-19/2)	ACB	1) 100 mL (C-19/2) 2) 100 mL (C-31)	SPFB SPFB	20 Ampoules	PB	40 Ampoules	PB	Pack size 1s 2 boxes	FP	10 Vials
	Juslin N	50 mL (C-19/2)	ACB	1) 100 mL (C-19/2) 2) 100 mL (C-31)	SPFB SPFB	20 Vials	PB	40 Vials	PB	Pack size 1s 2 boxes	FP	N/A
17.	Juslin 30/70	50 mL (C-19/2)	ACB	1) 100 mL (C-19/2) 2) 100 mL (C-31)	SPFB SPFB	20 Vials	PB	40 Vials	PB	Pack size 1s 2 boxes	FP	N/A
18.	Vancolon 1.0 g injection	50 mL (C-19/2)	ACB	1) 100 mL (C-19/2) 2) 100 mL (C-31)	SPFB SPFB	–	–	–	–	–	–	–

continued

(continued)

S. No.	Product Name	Chemical Bulk (g/mL)	Container	Micro Bulk (g/mL)	Container	F. Chemical Bulk after Leak Test Autoclave Lyophilization	Container	Micro F. Bulk after Leak Test Autoclave Lyophilization	Container	Physical	Container	After Visual Inspection for Liquid-Borne Particulate Matter
19.	ENOXIRT 0.25 g IM	–	–	–	–	–	–	–	–	Pack size 1s 5 boxes	–	10 Vials
20.	ENOXIRT 0.25 g IV injection	–	–	–	–	–	–	–	–	Pack size 1s 5 boxes	–	10 Vials
21.	ENOXIRT 0.5 g IM	–	–	–	–	–	–	–	–	Pack size 1s 5 boxes	–	10 Vials
22.	ENOXIRT 0.5 g IV injection	–	–	–	–	–	–	–	–	Pack size 1s 5 boxes	–	10 Vials
23.	ENOXIRT 1 g IM	–	–	–	–	–	–	–	–	Pack size 1s 5 boxes	–	10 Vials
24.	ENOXIRT 1 g IV injection	–	–	–	–	–	–	–	–	Pack size 1s 5 boxes	–	10 Vials
25.	ENOXIRT 1 g IM/IV injection	–	–	–	–	–	–	–	–	Pack size 1s 5 boxes Pack size 10s 1 box	–	10 Vials
26.	NEGACEF 2 g IM/IV injection	–	–	–	–	–	–	–	–	Pack size 1s 5 boxes Pack size 10s 1 box	–	10 Vials

27.	Chlorohistol 10 mg/mL injection	50 mL	ACB	100 mL	SPFB	20 Ampoules	PB	40 Ampoules	PB	Pack size 50s 1 box	FP	20 Ampoules
28.	Chlorohistol 20 mg/2 mL injection	50 mL	ACB	100 mL	SPFB	20 Ampoules	PB	40 Ampoules	PB	Pack size 50s 1 box	FP	10 Ampoules
29.	Cimetag 2 mL injection	50 mL	ACB	100 mL	SPFB	20 Ampoules	PB	40 Ampoules	PB	Pack size 50s 1 box	FP	10 Ampoules
30.	Lovrak IV for 250 mg infusion	50 mL	ACB	100 mL	SPFB	20 Ampoules	PB	40 Ampoules	PB	Pack size 50s 1 box	FP	10 Vials
31.	Omedoc 40 mg IV injection	50 mL	ACB	100 mL	SPFB	20 Ampoules	PB	40 Ampoules	PB	Pack size 50s 1 box	FP	10 Vials
32.	Gastroprotect 2 mL injection	50 mL	ACB	100 mL	SPFB	20 Ampoules	PB	40 Ampoules	PB	Pack size 50s 1 box	FP	10 Ampoules

QAS-004

YOUR COMPANY VALIDATION STANDARD OPERATING PROCEDURE

SOP No. QAS-004.00 Effective date: mm/dd/yyyy
Approved by:

TITLE: **Sampling of Bulk Products for Laboratory Testing**

AUTHOR: ______________________________

Name/Title/Department

____

Signature/Date

CHECKED BY: ______________________________

Name/Title/Department

____

Signature/Date

APPROVED BY: ______________________________

Name/Title/Department

____

Signature/Date

REVISIONS:

No.	Section	Pages	Initials/Date

SOP No. QAS-004.00 Effective date: mm/dd/yyyy
Approved by:

SUBJECT: Sampling of Bulk Products for Laboratory Testing

23.1 PURPOSE

To provide written instructions for the sampling of bulk products.

23.2 RESPONSIBILITY

It is the responsibility of the QAI to follow up the procedure. The QA manager will be responsible for the SOP compliance.

23.3 PROCEDURE

23.3.1 Sampling of Bulk Products and Intermediate Products

Prerequisite: The QAI will make sure that the container is clean, sterile, dry, and pyrogen free (as the case may be).

23.3.2 Dry Production: Tablets and Capsule Bulk Sampling

After release as per SOP QAS-000 by quality assurance, the QAI will label plastic container/plastic bags with the sampling label and will raise a sampling report by mentioning the product's name, code and batch number, and stage of sampling (Attachment I).

At the commencement of the compression/encapsulation, the QAI will send a sample of 50 tablets/capsules with bulk release form for analysis/dissolution test from every batch.

The QAI will collect tablets/capsules in plastic container/plastic bag such that the representative sample of the whole batch is collected and the sample size is as per the quantities required for testing of each product.

Container/plastic bags will be sent to the quality control administration for analysis and bulk release.

23.3.2.1 Powder Prosuspension

The QAI will collect the samples after blending (or as mentioned in the product specification) in plastic bag or bottles with the help of a sampling thief.

The QAI will prepare a sampling report making sure that the samples are properly labeled.

23.3.2.2 Sampling after Sealing/Coating

The information for sampling will be provided by the production department on "sampling advice" at the time of completion of sealing and coating.

The QAI will sample and prepare the sampling report, making sure that the sample is properly labeled and the report is duly filled in.

23.3.3 Sampling of Soft Products

At the completion of bulk manufacturing, the production department will inform the QAI to collect samples. The QAI will collect samples in glass bottles or plastic container (cream and ointment) and will send the sample along with the sampling report to the in-process laboratory at the packaging side.

23.3.3.1 Exception

Bulk sampling of suspension products for microbiological testing shall be carried out from filling lines.

SOP No. QAS-004.00 Effective date: mm/dd/yyyy
Approved by:

23.3.4 Sampling of Injectables

At the completion of bulk manufacturing, the QAI will be informed by the production operator to collect the samples for microbiological test. The QAI will collect the samples in sterile pyrogen-free bottles for microbiological tests and in clean amber glass bottles for chemical analyses. The QAI will then label the bottles appropriately with the product's name, batch/lot number, and code number and will send them to the QC laboratory along with bulk release form (Attachment II). The QAI will register the samples in the bulk sampling register of QA.

23.3.5 For Sterility Testing

In order to represent the entire batch and processing conditions, samples should be taken

- At the beginning, middle, and end of the aseptic processing operations
- In conjunction with processing interventions or excursions

Processing excursions or interventions could be but not limited to the following:

Excursions to the time limits (durations) of

- The period between the start of the bulk product compounding and its sterilization
- Filtration processes
- Product exposure while on the processing line
- Storage of the sterilized equipment, containers, and closure

Interventions that occur:

- Maintenance
- Stoppages
- Equipment adjustments

The sample for the sterility test will be divided over the filling process plus samples in conjunction with the intervention and excursion. The QAI will document the sampling time and reasons on the control chart and will mark the ampoules to correlate accordingly.

23.3.6 Semifinished Sampling for Chemical

The samples after terminal sterilization, or cold leak test of the ampoules after capping of vials, are collected in a polybag. The polybag should be labeled appropriately with the product's name, batch/lot number, and code number. The area QAI will send the samples along with the bulk release form.

23.3.7 For Liquid-Borne Particulate Matter

Sampling for liquid-borne particulate matter is carried out after visual inspection. Samples are collected in a labeled polybag. The QAI will make entries in a sampling register of QA and will send the samples to QC with the bulk release form (Attachment III).

23.3.8 Resampling Request in Case of Repeat Testing

The request will come through production along with "sampling advice" mentioning the reason for resampling, for example, whether a cumulative sample or individual container samples are required.

SOP No. QAS-004.00 Effective date: mm/dd/yyyy
Approved by:

Bulk release form (completely filled and signed) should always accompany the sample to QC administration, mentioning the testing required at the stage of sampling.

The QAI makes sure that

1. Containers for sampling are clean and dry or sterile and pyrogen free.
2. Bulk product and container are properly labeled.
3. Documents are as per "checklist for batch documents for bulk product" (Attachment II) for injection along with required documents to QC.

23.4 REASON FOR REVISION

First time issued for your company.

23.5 ATTACHMENT FORMS

ABC PHARMACEUTICAL COMPANY

LEVEL-3
SOP No.: QAS-004
Issued on:
Revision No.:
Initiator name:

QUALITY ASSURANCE DEPARTMENT

Attachment I

SAMPLING REPORT

Product name: ____________________

Code No.: ____________________

Batch No.: ____________________

Process (stage of sampling): ____________________

Number of containers (in the batch): ____________________

Number of containers sampled: ____________________

Sample quantity: ____________________

Sampled by: ____________________

Date: ____________________

- Visual inspection of bulk product and containers ____________________
- Labeling requirement satisfactory or not ____________________
- Any other information ____________________
- Any documents attached with the sample: production documents/MFM requisitions ____________________
- QA in-process documents ____________________ ____________________
- Remarks____________________ ____________________ ____________________

QAI/QAO

(Signature)

ABC PHARMACEUTICAL COMPANY

LEVEL-3
SOP No.: QAS-004
Issued on:
Revision No.:
Initiator name:

Quality Assurance Department

LEVEL 4
SOP No.: QAS-004

Attachment II
BULK RELEASE

Product name: ____________ Batch No.: ____________ Code No.: ____________

Manufacturing date: ____________ Exp. date: ____________

Manufacturing		
– Ampoules	☐ Content	☐ Physical parameters
– Capsules	☐ Dissolution	☐ Content (assay)
– Cream	☐ Content	
– Drops	☐ Content	
– Ointment	☐ Content	
– PPS	☐ Content	
– Suppositories	☐ Content	☐ Content uniformity
– Suspensions	☐ Content	☐ Content uniformity
– Syrups	☐ Content	☐ Physical parameters (pH, refractive index, etc.)
– Tablets/caplets	☐ Granules	☐ Core ☐ Coated ☐ Content uniformity
– Vials	☐ Content	
– Others	☐ Content	

Remarks* ____
____
____

Date: ____________ Time: ____________ Name & Signature: ____________

QC No.: ________ Date: ________ Time: ________ Signature: ____________

Physical Testing:

____
____
____
____

Analyzed by: ____________________ Date: ____________ Time: ____________

ABC PHARMACEUTICAL COMPANY

LEVEL-3
SOP No.: QAS-004
Issued on:
Revision No.:
Initiator name:

Chemical Testing:

____
____
____
____

Analyzed by: ________________ Date: ________ Time: ______

☐ Released for packaging ☐ Not released for packaging

Checked by: Manager QC **Approved by: QC Director**

*Other specific test needed

ABC PHARMACEUTICAL COMPANY

LEVEL 3
SOP No.: QAS-004
Issued on:
Revision No.:
Initiator name:

Attachment III

SAMPLING STAGE AND SAMPLE AMOUNT FOR DIFFERENT PRODUCTS

S. No.	Product Name	Sampled Quantity and Type of Container							
		Bulk (gram)	Chem. Bulk	Container	Physical		Microbiological		
					Bulk	Finished	Bulk	Finished	Container
Tablets/Syrup/Ointments & Creams/Granules/Injectables									

QAS-005

YOUR COMPANY STANDARD OPERATING PROCEDURE

SOP No. QAS-005.00 Effective date: mm/dd/yyyy
Approved by:

TITLE: **Product Recall Procedure**

AUTHOR: ______________________________

Name/Title/Department

____

Signature/Date

CHECKED BY: ______________________________

Name/Title/Department

____

Signature/Date

APPROVED BY: ______________________________

Name/Title/Department

____

Signature/Date

REVISIONS:

No.	Section	Pages	Initials/Date

SOP No. QAS-005.00 Effective date: mm/dd/yyyy
Approved by:

SUBJECT: Product Recall Procedure

24.1 PURPOSE

To provide written instructions to be followed for prompt action in the event of a minor, major, and critical complaint which is suspected or known to be a defective or hazardous product.

24.2 RESPONSIBILITY

It is the responsibility of directors QA and QC, and the QA manager to identify the basis of recall in collaboration with the product recall committee. The QA director is responsible for the complete recall operation.

24.3 PROCEDURE

1. The objective is to establish a procedure that will enable prompt action to be taken to recall a product that is suspected or known to be defective or hazardous. If recalled, the product will be kept in a segregated area.
2. The responsibility for identifying the possible need to recall a product also rests with any member of the management whose knowledge and/or technical training will enable him to identify that a potentially defective or hazardous product has been supplied to the market or to a customer. This manager has an immediate obligation to inform the general manager of his opinions in this matter.
3. The responsibility for initiating a product recall rests with the general manager. To assist him in discharging this responsibility, he will consider the advice offered by the *product recall committee.*
4. The members of the *product recall committee* are as follows:
   - QA director
   - Production director
   - Materials director
   - Marketing director or representative
   - QC director
   - QA manager
   - R&D director
   - Any other employee considered resourceful by the committee
   - Legal advisor of the company, if considered applicable
   - The QA director will identify the scope and nature of the defect or hazard
   - The QC director will check the scope and nature of the defect or hazard
   - The QA manager will verify the scope and nature of the defect or hazard
   - The marketing director will identify the possible market/consumer penetration of the defective or hazardous product and collect information on relevant forms
5. In the absence of the general manager of the company, the QA director will be responsible for initiating a product recall.
6. Soon after a complaint is received through any source or management level, it is the responsibility of the receiver of the complaint to pass the information immediately to the QA director, QA manager, or QC director, whichever is applicable. The marketing director

SOP No. QAS-005.00 Effective date: mm/dd/yyyy
Approved by:

will be responsible for locating and bringing the defective sample/s of the product to the QC department to trigger the investigative analysis on an immediate basis and for other necessary actions.

7. Meanwhile, the QA director will immediately pass on the information to the general manager and organize a meeting with the product recall committee to identify the type of recall. Refer to Annexure II for recall classification.
8. After a tentative conclusion of the type of recall, the QA manager will make entries on Annexure III (Product Recall Investigation Form II) before the final meeting.
9. Immediately after the tentative disposition, the QA director informs the general manager and advises an immediate provisional freezing of the relevant stocks at the factory store. Wholesalers, institutions, pharmacies, distributors, and institutions outside the country will also be instructed to freeze distribution till further instructions.
10. In the absence of the QA director, the QA manager and the QC director will be responsible for the recall operation.
11. Meanwhile, the general manager will create a team of marketing personnel to start collecting more information to establish facts on Annexure I (Product Recall Investigation Form I) for legal and technical actions.
12. For corrective actions, refer to Annexure IV.
13. Immediately after the advice offered by the *product recall committee*, the QA director will call a meeting of the managers with the general manager for the final conclusion and corrective actions.
14. The legal advisor of the company may also be contacted for his opinion.
15. The meeting objective will be to reassess the depth of recall or other remedial action in light of the discussion.
16. If considered appropriate, the general manager will also seek advice from the ministry of health.
17. The possible decisions may be
    i. No recall required.
    ii. Recall from stores and distribution agents.
    iii. Complete recall up to direct users (the general manager may decide to use radio and television network, if necessary).
18. The ministry of health and all competent authorities of all countries to which the product/s may have been distributed will be informed promptly if the product/s are to be recalled.
19. If the decision is no recall, the marketing director will instruct all the distribution centers to release the stocks, if not replaced by a new batch by that time.
20. The final meeting of the drug recall committee shall be held at the end to review the case and the steps taken to avoid a similar complaint in future.
21. If appropriate, the committee shall also decide regarding the appropriate compensation scheme to be followed.

24.4 DOCUMENTATION

Annexure I	Product Recall Investigation Form I
Annexure II	Recall Classification
Annexure III	Product Recall Investigation Form II
Annexure IV	Corrective Action Form

SOP No. QAS-005.00 Effective date: mm/dd/yyyy
Approved by:

Note:

- The documentation related to product recall for US-marketed product (ANDA) will be maintained by the client in the country of sales. A copy should be marked to ABC Pharmaceutical Company.
- ABC Pharmaceutical Company will perform and maintain the investigation records related to the product material reviewing, manufacturing, packaging, testing, and shipping.

24.5 REASON FOR REVISION

First time issued for the ABC Pharmaceutical Company.

24.6 ATTACHMENT FORMS

ABC PHARMACEUTICAL COMPANY

LEVEL-3
SOP No.: QAS-005
Issued on:
Revision No.:
Initiator name:

ANNEXURE I
Product Recall
Investigation Form I

Problem received on ____________________ (by) quality assurance director

Time:

1. Product name : ____________________ Strength: ______________
2. Code No. : ____________________
3. Batch No. : ____________________
4. Mfg. date : ____________________ Exp. Date: ______________
5. Pack size : ____________________
6. Problem origin:
   a. Department of health: ____________________
   b. Medical practitioner: ____________________
   c. Hospital pharmacist: ____________________
   d. Retail pharmacist: ____________________
   e. Distributor: ____________________
   f. End user (patient): ____________________
   g. Field: ____________________
   h. Other: ____________________
7. Exact nature of problem: ____________________
   ____
8. Other details: ____________________
   a. Name of the problem reporter: ____________________
   b. Address of the problem reporter: ____________________
   ____________________ Phone No.: ______________

ABC PHARMACEUTICAL COMPANY

LEVEL-3
SOP No.: QAS-005
Issued on:
Revision No.:
Initiator name:

c. Name of the institution where the reaction occurred:

______________________________ Phone No.: __________

d. Name of persons already notified, e.g., health department:

______________________________ Phone No.: __________

e. Consultation doctor's name and address: ______________________________

______________________________ Phone No.: __________

9. Product prescribed for treatment of: ______________________________

If Patient/Patients

10. No. of patients involved: ______________________________

11. Name of patients — Date effected — Symptoms

	Name of patients	Date effected	Symptoms
1.			
2.			
3.			
4.			
5.			

12. Regular dosage/dosage administered

	Patient's name	Usage	Dosage administered
1.			
2.			
3.			
4.			
5.			

ABC PHARMACEUTICAL COMPANY

LEVEL-3
SOP No.: QAS-005
Issued on:
Revision No.:
Initiator name:

13. Names/dosage of other drugs used concurrently

	Patient's name	Other drugs/drugs used	Dosage
1.	__________	__________	__________
2.	__________	__________	__________
3.	__________	__________	__________
4.	__________	__________	__________
5.	__________	__________	__________

14. Response of patient to other drugs

	Patient's name	Other drug/drugs used	Adverse reactions (If yes, symptoms)
1.	__________	__________	__________
2.	__________	__________	__________
3.	__________	__________	__________
4.	__________	__________	__________
5.	__________	__________	__________

ABC PHARMACEUTICAL COMPANY

LEVEL-3
SOP No.: QAS-005
Issued on:
Revision No.:
Initiator name:

ANNEXURE II

Recall Classification

Class I: Critical (Possible Reasons)
Product's defect that may result in serious health consequences or death.

Mix-Ups (Partial Batches)

1. Active ingredients
2. Inactive ingredients (e.g., incompatible excipient yielding pharmacologically active impurities)
3. Semifinished products (e.g., sustained released product with one coating only)
4. Printed packaging components (e.g., 5 mg strength of some potent drug filled in a printed packaging material of 1 mg)

Mistaken Identity (Whole Batch)

1. Active ingredients (e.g., life-saving drug (replaced or mixing))
2. Semifinished products (e.g., mix of a potent drug before coating with some nonpotent drug having similar finishing)
3. Printed packaging components (improper strength of a life-saving medicine packed in wrong packaging material)

Impurities

1. Physiologically active ingredients
2. Pathogenic microorganism
3. Turbidity, precipitation in injectables

Deviations/Changes

1. Active ingredients missing
2. Incorrect dosing of active ingredient that might result in serious adverse health consequences or death
3. Product decomposition that might result in serious adverse health consequences or death
4. Change in packaging components that might result in serious adverse health consequences or death
5. Change in active ingredient release that might result in serious adverse health consequences or death
6. Unexplainable changes in effectiveness that might result in serious adverse health consequences or death
7. Counterfeit

ABC PHARMACEUTICAL COMPANY

LEVEL-3
SOP No.: QAS-005
Issued on:
Revision No.:
Initiator name:

Decision: The product shall be completely recalled from all the distribution points, including individual consumers in a minimal time. Also use telecommunication resources.

Class II: Major (Possible Reasons)

Product defects which may not cause adverse health consequences but will impair efficacy or usefulness.

The intensity of the problem shall be evaluated using military standards 105 E. Tightened sampling plan shall be used to reach the final conclusion.

Mix-Ups

1. Inactive ingredients (e.g., mix-up of compatible excipient, only affecting the therapeutic values)

   The sampling and testing shall be performed at 0.65% acceptable quality level (AQL) single sampling using level S-3; otherwise the batch shall be recalled from the market.

Impurities

1. Nonpathogenic microorganisms in excess of tolerance limits
2. Excessive amounts of physiologically harmless materials

Deviations/Changes

1. Perceptible changes in appearance, odor, color, taste, physicals, etc. of the product (1)
2. Deviations in packaging components (e.g., only leaflet change) (1)
3. Expiration date (1)
4. Product that would impair usability (e.g., broken tablets)
5. Defects in packaging components that would impair usability (e.g., leaks, pinholes, functional defects)
6. Missing labels
7. Empty packages
8. Underfilling, in terms of volume, weight, or quantity, and overweight (2)
9. Labels peeled off cartons, missing insert (2)

Remarks

1. Major in the sense that it could result in misinterpretation and uncertainty on the part of the consumer.
2. The extent of the defect is to be considered here. Under certain circumstances, it could be classified as a Class III defect.

Class III: Minor (Possible Reasons)

Product's defects which will cause no adverse health consequences and/or effect on product efficacy.

ABC PHARMACEUTICAL COMPANY

LEVEL-3
SOP No.: QAS-005
Issued on:
Revision No.:
Initiator name:

Deviations/Changes

1. Minor deviations from product standards (e.g., in terms of color, odor, taste, shape, or occurrence of spots)
2. Standard deviations from packaging component standards (e.g., color, occasional illegible printing of lot number, inconsequential misprints or misprints such as outdated registration numbers)
3. Packaging defects such as crooked labels and rubbed-off print

ABC PHARMACEUTICAL COMPANY

LEVEL-3
SOP No.: QAS-005
Issued on:
Revision No.:
Initiator name:

ANNEXURE III

Product Recall

Investigation Form II

1. Responsibility: To be filled by all the members of the product recall committee to arrive at a final conclusion regarding the type of recall.
2. Classification of problem

   Class I: Critical ______________________

   Class II: Major ______________________

   Class III: Minor ______________________

3. Comments on analysis of recall product and retention samples by QC director:

   ____

   ____

   ____

4. Comments on batch documents reviewed by QA manager and production director:

   ____

   ____

5. Other batches which contain reworks of the defective batch (production director):

   ____

   ____

   ____

6. Location and distribution of stock from the batch (marketing director):

   ____

7. Comments of marketing director: ______________________

   ____

   ____

ABC PHARMACEUTICAL COMPANY

LEVEL-3
SOP No.: QAS-005
Issued on:
Revision No.:
Initiator name:

8. Comments of packaging director: ______________________________

____

____

9. Comments of R&D director: ______________________________

____

____

10. Comments of director quality assurance: ______________________________

____

____

11. Corrective actions: ______________________________

____

____

ABC PHARMACEUTICAL COMPANY

LEVEL-3
SOP No.: QAS-005
Issued on:
Revision No.:
Initiator name:

ANNEXURE IV

Product Recall

Agenda for Final Conclusion and Corrective Actions

1. *Responsibility*: The input from Annexure III will be transferred to this form. The quality affairs director will be responsible for the following procedure.
2. *Problem classification*:

Manager	Problem Classification	Comments
1. QA director		______ ______ ______
2. QC director		______ ______ ______
3. QA manager		______ ______ ______
4. Marketing director		______ ______ ______
5. Production director		______ ______ ______
6. Packaging and materials handling director		______ ______ ______
7. PDL director		______ ______ ______

ABC PHARMACEUTICAL COMPANY

LEVEL-3
SOP No.: QAS-005
Issued on:
Revision No.:
Initiator name:

3. Comments of legal advisor (if contacted): ______________________

____

____

4. Suggestion of ministry of health (if contacted by the general manager):

____

____

5. Conclusion by the general manager:

Drug recall class: ______________________

Reasons: ______________________

____

____

6. Corrective actions: ______________________

____

____

____

QAS-006

YOUR COMPANY STANDARD OPERATING PROCEDURE

SOP No. QAS-006.00 Effective date: mm/dd/yyyy
Approved by:

TITLE: **Cream/Ointment In-process Audit Checks**

AUTHOR: ______________________________

Name/Title/Department

____

Signature/Date

CHECKED BY: ______________________________

Name/Title/Department

____

Signature/Date

APPROVED BY: ______________________________

Name/Title/Department

____

Signature/Date

REVISIONS:

No.	Section	Pages	Initials/Date

SOP No. QAS-006.00 Effective date: mm/dd/yyyy
Approved by:

25.1 PURPOSE

This SOP outlines the stepwise procedure for auditing and confirming the fill weight and physical attributes of cream and ointment tubes and also to check all other requirements of the area.

25.2 RESPONSIBILITY

The QA inspector of the area and the QA manager are responsible to follow up the procedure whereas the QA director is responsible for the SOP compliance.

25.3 PROCEDURE

In-process audit checks to be carried out on an hourly basis. The following are the checks to be carried out during the cream/ointment filling operation. This procedure is divided into two portions.

- A. Checks to be carried out to assure the system and process
  - A.1 Equipment and area clearance checks
  - A.2 Line release
- B. Checks to be carried out during filling operations
  - B.1 Physical characteristics of container closures and the products
  - B.2 Weight variation checks
  - B.3 Leak test

A.1 Equipment and area clearance checks

- A.1.1 During the first round of the area, the QA inspector will check the presence of equipment and area clearance will be granted by the supervisor.
- A.1.2 Check that all packers/operators are properly dressed and their hair are properly covered.
- A.1.3 Check that the operators are wearing masks, gloves, and safety glasses.
- A.1.4 Check that no other batch is present in the area except for the batch that is scheduled for filling.
- A.1.5 Ensure that only the respective batch documents are present on the line.
- A.1.6 Tally the product's name and batch number on line clearance with product-related documents.
- A.1.7 If all the GMP (good manufacturing practice) requirements and clearance requirements are found satisfactory, then allow the process to start and sign off the line clearance sheet. If any deviation is observed, write in the REMARKS column of the in-process chart (X & R chart).

A.2 Line release
Once the machine is set, the supervisor will call the QA inspector to check

1. Average weight (filled)
2. Weight variation
3. Leakage
4. Physical attributes
5. Batch engraving on the tubes and boxes
6. Completed packs as per product specification

If found satisfactory, the QA inspector will issue the release label.

SOP No. QAS-006.00 Effective date: mm/dd/yyyy
Approved by:

B.1 During the filling operation, checks are carried out for the physical attributes of the filled packs.

Procedure

1. Check cleanliness of the empty tubes fitted with caps.
2. Collect 32 tubes from the filled units on an hourly basis and check the defects of tubes, if any, such as deshaped tubes, pin holes, dented tubes, improper crimping, and so on and for broken or cracked caps. If any defective tube is found, inform the supervisor for corrective action.
3. Enter the findings in a relevant control chart.

B.2 Weight variation checks
Confirm that the machine has been set in limits before the operation by a line boss and machine operator as per the written procedure.

Procedure

1. Once the machine has been set within limits and the process has been allowed to run, start collecting the samples on an hourly basis.
2. Weigh the empty tubes on a top pan balance to get an average weight of empty tubes to be used in the batch.
3. Calculate the gross limits and enter them on the chart.
4. Get the filled tubes directly from the machine on an hourly basis.
5. Weigh the filled tube that should be within limits.
6. Record the results on a control chart and decide disposition as per attached appendix or product in-process specification (packaging).

B.3 Leak test
During cream/ointment filling operation carry out the leak test at the time of setting or beginning of the process and at least once in a shift.

Procedure

1. Collect 10 tubes randomly after filling
2. Place the tubes in an oven
3. Close the oven and set the temperature at 60–65°C
4. Keep for 15–20 min or until the contents melt

Action Limit

1. If any leakage is observed out of 10 units, check for reasons of leakage.
2. If leakage happened due to crimping and sealing, cracked cap, without wads, pin holes, and so on, inform the packaging supervisor for rectifying the anomaly and sorting of already filled or packed units since the last check.

25.4 REASON FOR REVISION

First time issued for the ABC Pharmaceutical Company.

25.5 ATTACHMENT FORM

ABC PHARMACEUTICAL COMPANY

LEVEL 3
SOP No.:
Issued on:
Revision No.:
Initiator name:

ANNEXURE I

DEFECTS (ATTRIBUTES TO BE CHECKED)

From filling point
Group I
Clarity
Less count (tablet & capsule)
Less fill check (liquid & granules)
Less removal torque
Chipped brim
Broken plastic cap
Cracked container
Deshaped tubes, dented tubes
Pinholes
Empty unit
Improper cutting
Pitting
Cracking
Edges
Cavity
Hole

Group II
Liquid spills
P.P. caps broken bridges
Cutmarks on P.P. caps
Improper crimping/sealing
Leak test

From packing
Critical defects
Wrong B. No./expiry/mfg.
Wrong special stamping
Empty unit
Foreign material
Incorrect No. of unit

Major defects
Wrong commodity
Missing component
Illegible B. No./mfg./exp.
Others

Minor defects
Open label
Crooked label
Torn label
Excessive shrinkage
Smeared overprinting
D-shaped cartons
Excessive gumming
Double labels
Smeared printing
Improper shrink wrap

ABC PHARMACEUTICAL COMPANY

LEVEL 3
SOP No.:
Issued on:
Revision No.:
Initiator name:

OINTMENT & CREAM

Pack Size (g)	Adjustment Limits (Preset) (g)	In-process Limits (g)
10	10–10.5	9.5–10.5
15	15–15.5	14.4–15.6
20	20–20.5	19.2–20.8
25	25–25.5	24.2–26.0
30	30–31.0	29.2–31.0
50	50–52.0	48.5–52.0
75	73.5–76.5	72.8–77.3
500	500–525	500–525
		(No single jar should contain less than 475 g)

Minimum fill

- The average net content of the 10 containers is not less than the labeled amount, and the net content of any single container is NLT 90% of the labeled amount where the labeled amount is 60 g or 60 mL or less, or NLT 95% of the labeled amount where the labeled amount is more than 60 g or 60 mL but not more than 150 g or 150 mL.

QAS-007

YOUR COMPANY STANDARD OPERATING PROCEDURE

SOP No. QAS-007.00 Effective date: mm/dd/yyyy
Approved by:

TITLE: **Inspection of Physical Attributes/Characteristics of Coated/Uncoated Tablets**

AUTHOR: ______________________________

Name/Title/Department

____

Signature/Date

CHECKED BY: ______________________________

Name/Title/Department

____

Signature/Date

APPROVED BY: ______________________________

Name/Title/Department

____

Signature/Date

REVISIONS:

No.	Section	Pages	Initials/Date

SOP No. QAS-007.00 Effective date: mm/dd/yyyy
Approved by:

26.1 PURPOSE

The purpose of this SOP is to outline a procedure for the final inspection of physical attributes of coated/uncoated tablets.

26.2 RESPONSIBILITY

The QAI is responsible for following the procedure. The QA manager will be responsible for SOP compliance.

26.3 PROCEDURE

The procedure to be followed if the QA is asked by the QC to carry out the test of physical inspection on the completion of a process prior to their checks.

The sampling and inspection procedure is as follows.

26.3.1 Sampling

a. After complete coating or compression, as the case may be, samples are to be collected by the QAI. The QAI has to examine each drum and ensure that it is:
   i. Correctly labeled
   ii. Contains the correct product
b. Tablets are to be sampled randomly from each drum according to batch size (or subbatch size) and Table I (Annexure I) based on ANSI/ASQC Z1.4.
c. The QAI should wear gloves and a mask during inspection.
d. Sampled tablets are to be spread over glass and observed thoroughly for defectives according to the acceptance plan.
e. The inspection has to be carried out as per the normal plan of ANSI/ASQC Z1.4 (see Annexure II).

26.3.2 Attributes Inspection: Coated Tablets

a. *Critical defects:* AQL 0.0%
   1. Foreign tablets
   2. Thick or thin tablets
   3. Foreign matter
   4. Faulty enteric coated tablet or controlled release coated tablets.
      a. Chipped coating
      b. Capped
      c. Sticking together
      d. Blistered coating
b. *Major defects:* AQL 1.0%
   Broken or cracked
   Capped
   Color variation
   Incomplete coating
c. *Minor defects:* AQL 2.5%
   Small specks

SOP No. QAS-007.00 Effective date: mm/dd/yyyy
Approved by:

Corrugation
Small chips
Incomplete coating
Bumpy rough coating
Broken punch marks
Sticking to tablet punch (slight)

26.3.3 Attributes Inspection: Uncoated Tablets

A. *Critical defects:* AQL 0.0%
Foreign tablets
B. *Major defects:* AQL 1.0%
Capping
Broken
Picking
Dirty
Large specks
C. *Minor defects:* AQL 2.5%
Small specks
Porous surface

26.3.4 Definition of Defects Related to Coated Tablets

A. *Critical defects*
Foreign tablets: any tablet other than the actual product.

B. *Major defects*

Broken	:	Complete breaking of tablets into two or more pieces.
Chipped	:	Removal of part of coating, i.e., core is visible.
Capped	:	A tablet showing separation of a part of it and the core of tablet becomes visible. Such a tablet may or may not break into two parts when shaken in the loose covered palm of hands.
Sticking	:	Tablets stick together in such a way that if separated they may break and the core becomes visible.
Blistered	:	A thin or thick bubble on the surface of tablets due to bulging of the coated solution.
Color	:	The color of the coating not meeting the accepted variation of color. Sample variation limit indicates tolerance.
Incomplete coating	:	Core visible partly at places due to inadequate coat level.

C. *Minor defects*

Small specks	:	Small visible foreign material on surface.
Corrugation	:	Unevenly applied coating, having a wavy or lumpy appearance. Sample limit indicates tolerance.
Small chips	:	Removal of part of tablet and core is not visible.
Incomplete coating	:	Subcoat visible at places particularly at the edges due to inadequate coat level.

SOP No. QAS-007.00 Effective date: mm/dd/yyyy
Approved by:

26.3.5 Definitions of Defects Related to Uncoated Tablets

Critical defects
Foreign tablets: Any tablet other than the actual product.

Major defects

A. Capping	:	Any tablet showing separation of a part of tablet partly or throughout its horizontal axis. Such tablets may or may not break into two parts.
B. Broken	:	Any tablet broken into two or more pieces.
C. Picking	:	Any tablet having depression of 2 mm or more on its surface.
D. Dirty	:	Any tablet with a spot of different color (not due to entrapped specks) or oil spots from machine.
E. Large	:	Any tablet with entrapped speck or foreign material of 2 mm or more in size.

Minor defects

A. Picking	:	A depression of less than 2 mm in size.
B. Small specks	:	Specks of less than 2 mm in size.
C. Porous surface	:	Unsmooth surface, resulting in loss of shine.

26.3.6 Definition of "Defective"

a. *Critical defective*: Critical defective by virtue of having one or more critical defects. Additionally a critical defective can have major or minor defects as well.
b. *Major defective*: By virtue of having one or more major defects. Additionally, a major defective can have minor defects but no critical defects.
c. *Minor defective*: By virtue of having one or more minor defects. A minor defective does not have major or critical defects.

26.3.7 Documentation

The findings/results will be entered in the format for physical attributes of coated or core tablets. The disposition will be decided on the basis of the acceptance plan attached.

a. If the sample is having defectives less than or equal to the acceptance number, then the lot is accepted.
b. If the sample is having defectives equal to or more than the rejected, then the lot is rejected.
c. Whereas if defectives are more than the acceptance number but less than the rejected number, then a second sample of the same size is taken and is inspected and the disposition is decided on the cumulative sample (the first and the second).

26.4 ATTACHMENT FORMS

ABC PHARMACEUTICAL COMPANY

LEVEL 3
SOP No.:
Issued on:
Revision No.:
Initiator name:

ANNEXURE I

ACCEPTANCE PLAN A—DOUBLE SAMPLING

NORMAL INSPECTION MIL. ANSI/ASQC Z1.4

				ACCEPTANCE QUALITY LEVEL					
				Critical AQL 0.0%		Major AQL 1.0%		Minor AQL 2.5%	
Sample Size Letter	Sample	Sample Size	Cumulative Sample Size	Acc.	Rej.	Acc.	Rej.	Acc.	Rej.
L	First	125	125	0	1	2	5	5	9
	Second	125	250	0	1	6	7	12	13
M	First	200	400	0	1	3	7	7	11
	Second	200	400	0	1	8	9	18	19
N	First	315	315	0	1	5	9	11	16
	Second	315	630	0	1	12	13	26	27

Names :
Acc. : Acceptance number
Rej. . Rejection number
* If acceptance number exceeded on the first sample, inspect another sample.
* If rejection number exceeded on the first sample, no second sample is taken.

ABC PHARMACEUTICAL COMPANY

ANNEXURE II LEVEL 3

INSPECTION OF PHYSICAL ATTRIBUTES
TABLET (CORE)

Product name: ____________ Batch size: ____________

Code No.: ____________ Sample size: ____________

Batch No.: ____________ Description: ____________

DEFECTS			BATCH SIZE	SAMPLE	ACCEPTANCE	
CRITICAL				**SIZE**	**AQL**	**0.0%**
Foreign product/odor					ACC.	REJ.
Gross foreign matter			35,001 to 150,000	125	0	1
Thick & thin tablet				125	0	1
Other defects			150,001 to 500,000	200	0	1
				200	0	1
Other defects			500,001 & over	315	0	1
Total critical defects / Sample size				315	0	1
MAJOR					**AQL**	**1%**
Capping					ACC.	REJ.
Broken			35,001 to 150,000	125	2	5
Dirty (specks/particles)				125	6	7
Picking (large)			150,001 to 500,000	200	3	7
Color variation (major)				200	8	9
Sticking			500,001 & over	315	5	9
Other defects				315	12	13
Total critical defects / Sample size						
MINOR					**AQL**	**2.5%**
Chipping					ACC.	REJ.
Picking (small)			35,001 to 150,000	125	5	9
Small specks				125	12	13
Pitting (porous surface)			150,001 to 500,000	200	7	11
Sticking (small)				200	18	19
Other defects			500,001 & over	315	11	16
Total critical defects / Sample size				315	26	27
Disposition:			Conforms Doesn't conform ____________			

Remarks:

____

____

____

____
Analyst

ABC PHARMACEUTICAL COMPANY

LEVEL 3

INSPECTION OF PHYSICAL ATTRIBUTES TABLET (COATED)

Product name: ______________ Batch size: ______________

Code No.: ______________ Sample size: ______________

Batch No.: ______________ Description: ______________

DEFECTS			BATCH SIZE	SAMPLE SIZE	ACCEPTANCE	
CRITICAL					**AQL**	**0.0%**
Foreign product/odor					ACC.	REJ.
Gross foreign matter			35,001 to 150,000	125	0	1
Faulty enteric coating				125	0	1
Thick & thin tablet			150,001 to 500,000	200	0	1
Other defects				200	0	1
Total critical defects / Sample size			500,001 & over	315	0	1
				315	0	1
MAJOR					**AQL**	**1%**
Capping					ACC.	REJ.
Broken			35,001 to 150,000	125	2	5
Dirty (specks/particles)				125	6	7
Picking (large)			150,001 to 500,000	200	3	7
Color variation (major)				200	8	9
Incomplete coating			500,001 & over	315	5	9
Other defects				315	12	13
Total critical defects / Sample size						
MINOR					**AQL**	**2.5%**
Chipping					ACC.	REJ.
Picking (small)			35,001 to 150,000	125	5	9
Small specks				125	12	13
Pitting (porous surface)			150,001 to 500,000	200	7	11
Corrugation				200	18	19
Blistered coating			500,001 & over	315	11	16
Other defects				315	26	27
Total critical defects / Sample size						
Disposition:			Conforms Doesn't conform	______________		

Remarks:

____

____

____

____

Analyst

QAS-008

YOUR COMPANY STANDARD OPERATING PROCEDURE

SOP No. QAS-008.00 Effective date: mm/dd/yyyy
Approved by:

TITLE: **Powder In-process Filling Audit Checks**

AUTHOR: ______________________________

Name/Title/Department

____

Signature/Date

CHECKED BY: ______________________________

Name/Title/Department

____

Signature/Date

APPROVED BY: ______________________________

Name/Title/Department

____

Signature/Date

REVISIONS:

No.	Section	Pages	Initials/Date

SOP No. QAS-008.00 Effective date: mm/dd/yyyy
Approved by:

27.1 PURPOSE

This SOP outlines the stepwise procedure for auditing and confirming the fill weight and attribute characteristics of containers, closures, and product and also to check other requirements of the area.

27.2 RESPONSIBILITY

The area QAI and officer will be responsible for following the procedure. The QA manager will be responsible for systems compliance.

27.3 PROCEDURE

In-process audit checks to be carried out on an hourly basis. This procedure is divided into two portions.

A. Audit checks to be carried out for ensuring the systems and the process
  A.1 Equipment and area clearance checks
  A.2 Humidity and temperature checks

B. Audit checks to be carried out during filling operations
  B.1 Attribute characteristics of containers, closures, and the product
  B.2 Fill weight checks
  B.3 Removal torque checks

A.1 Equipment and area clearance checks

A.1.1 On the first round of the area, the QAI will check the line clearance granted by the supervisor.
A.1.2 Confirm that the area and the equipment are clean.
A.1.3 Confirm the cleanliness and proper labeling on the bulk container in which powder is kept.
A.1.4 Check whether the operator is wearing a mask, gloves, and clogs and the hair is properly covered.
A.1.5 Tally the product name and the batch number on the bulk container with the Batch Packaging Record and Master Packaging Instructions requisition.
A.1.6 Check that all the commodities being used in the area are as per the Master Packaging Instructions requisition and no other commodities are present in the area.
A.1.7 If all the requirements are met, then sign on the line clearance for start-up release.
A.1.8 When the machine is set and the finished product characteristics are found to be as per specifications, then give the line release by issuing the line release label.

A.2 Humidity checks (by the psychrometer)

A.2.1 Check the relative humidity (RH) of the powder filling area once/shift and at the start of filling packaging.
A.2.2 RH should be within specified limits.
A.2.3 Temperature: 25°C + 2°C or as specified in the Master Packging Instructions
A.2.4 Record the RH and temperature on the control chart.

SOP No. QAS-008.00 Effective date: mm/dd/yyyy
Approved by:

B.1 Physical attributes of containers, closures, and product

B.1.1 Confirm that the bottles and caps which are used in the area are clean and no foreign particles are present in them.

B.1.2 Collect samples randomly from filled units as per the sampling and acceptance plan and check the physical characteristics of containers, closures, and product for any defects as follows:

i. Foreign particles in the granules
ii. Chipped brim
iii. Broken cap and cracked containers
iv. P.P. caps broken bridges, cut marks on P.P. caps

B.1.3 Confirm that the checks taken by the line supervisor in that period is as per the written procedure.

B.1.4 Enter the readings and deviations on the control chart for granules filling.

B.1.5 If audit checks cross the cumulative acceptance number, then inform the supervisor and ask to segregate the portion.

B.1.6 Collect samples again from the segregated portion and decide the disposition of the segregated lot. If it passes, then release the portion. In case of failure, quarantine the segregated lot and ask for 100% inspection.

Note: The segregated portion could be completely finished units or partially finished or both.

B.2 Fill weight checks

Confirm that the machine has been set within pre-set limits before operation by the line boss.

B.2.1 Once the machine has been set and the process has been allowed to run, collect the samples on an hourly basis.

B.2.2 Take one filled bottle from each nozzle of the running machine. If the machine has only one nozzle, at least three bottles should be taken for weight variation.

B.2.3 Completely empty them on the pan of the balance one by one making sure that no powder is left in the bottle.

B.2.4 Note the net weight of each individual bottle.

B.2.5 Record the results on the control chart for granules filling.

B.3 Removal torque checks

B.3.1 Collect samples randomly from filled and sealed bottles as per the sampling plan.

B.3.2 Confirm the removal torque on a torque tester.

B.3.3 Note the torque readings on the control chart for granules filling.

B.3.4 If audit checks cross the cumulative acceptance number, then inform the packaging supervisor and ask to segregate the portion.

B.3.5 Collect samples again and decide the disposition of the segregated lot. If it passes, then release the portion. In case of failure, quarantine the segregated lot and ask for 100% inspection.

Note: The segregated portion could be completely finished units or partially finished or both.

27.4 ATTACHMENT FORM

ABC PHARMACEUTICAL COMPANY

LEVEL 3
SOP No.:
Issued on:
Revision No.:
Initiator name:

ANNEXURE I

DEFECTS (PHYSICAL ATTRIBUTES TO BE CHECKED)

From filling point
Group I
Clarity
Less count (tablet & capsule)
Less fill check (liquid & granules)
Less removal torque
Chipped brim
Broken plastic cap
Cracked container
Deshaped tubes, dented tubes
Pinholes
Empty unit
Improper cutting
Pitting
Cracking
Edges
Cavity
Hole

Group II
Liquid spills
P.P. caps broken bridges
Cutmarks on P.P. caps
Improper crimping/sealing
Leak test

From packing
Critical defects
Wrong B. No./expiry/mfg.
Wrong special stamping
Empty unit
Foreign material
Incorrect No. of unit
Wrong commodity

Major defects
Missing component
Illegible B. No./mfg./exp.
Others

Minor defects
Open label
Crooked label
Torn label
Excessive shrinkage
Smeared overprinting
D-shaped cartons
Excessive gumming
Double labels
Smeared printing
Improper shrink wrap

ABC PHARMACEUTICAL COMPANY

LEVEL 3
SOP No.:
Issued on:
Revision No.:
Initiator name:

ANNEXURE II

LIMITS FOR FILLING GRANULES (Granules for Oral Suspension)

Code	Product Name	Package Size (mL)	Fill Wt/mpi (g)	Limit (g)	Average (g)	Target (g)	Range (g)
	Cephalosporin A 125 MG Granules	60	39.3	39.3–41.26	35.37	39.3	39.0–39.6
	Cephalosporin A 250 MG Granules	60	39.3	39.3–41.26	35.37	39.3	39.0–39.6
	Cephalosporin A 125 MG Granules	100	65.5	65.5–68.77	58.95	65.5	65.0–66.0
	Cephalosporin A 250 MG Granules	100	65.5	65.5–68.77	58.95	65.5	65.0–66.5
	Cephalosporin B 250 MG Granules	100	66	62.7–69.3	62.7	66.0	65.5–68.0
	Cephalosporin B 125 MG Granules	100	66	62.7–69.3	62.7	66.0	65.5–67.0
	Amoxycillin + Clavulanate A 250 MG Granules	100	65.0	64.5–67.5	66.0	65.0	64.5–66.0
	Amoxycillin + Clavulanate A 125 MG Granules	100	65.0	64.5–67.5	66.0	65.0	64.5–66.0
	Penicillin A 250 MG Granules	100	65.0	64.5–67.5 62.0–68.0	66.0 62.0	65.0	
	Penicillin A 125 MG Granules	100	65.0	64.5–67.5 62.0–68.0	66.0 62.0	65.0	

Code	Product Name	Package Size (mL)	Fill Wt/mpi (g)	Limit (g)	Average (g)	Target (g)	Range (g)
	Penicillin B 250 MG Granules	60	39.0	38.7–40.4 37.0–41	39.55 37.0	MPI (Not available)	—
	Penicillin B 125 MG Granules	60	39.0	38.7–40.4 37.0–41	39.55 37.0	MPI (Not available)	—
	(1) (2) Amoxycillin 250 MG/5 ML PPS	60	39.0	38.7–40.4	39.55	MPI (Not available)	—
	Cephalosporin 250 MG/5 ML PPS	100	62.04	62.04– 68.57	62.04	100 ML MPI (Not available)	—
	Cephalosporin 250 MG PPS	75	49	44.1–53.9	44.1	75 ML 49.0	48.5–49.5

QAS-009

YOUR COMPANY STANDARD OPERATING PROCEDURE

SOP No. QAS-009.00 Effective date: mm/dd/yyyy
Approved by:

TITLE: **Sealing Checks of Strips and Blisters**

AUTHOR: ______________________________

Name/Title/Department

____

Signature/Date

CHECKED BY: ______________________________

Name/Title/Department

____

Signature/Date

APPROVED BY: ______________________________

Name/Title/Department

____

Signature/Date

REVISIONS:

No.	Section	Pages	Initials/Date

SOP No. QAS-009.00 Effective date: mm/dd/yyyy
Approved by:

28.1 PURPOSE

This SOP outlines the procedure for correct strips/blisters, sealing checks, attribute characteristics of product, foil, and area requirements.

28.2 RESPONSIBILITY

The area QAI and officer will be responsible for implementing the procedure and the QA manager will be responsible for SOP compliance.

28.3 PROCEDURE

A. Audit Checks to be Performed for Line Clearance
At the start the QAI will check the following.

Note: Check all the aluminum foil rolls to ascertain that printing of the matter is on the correct side (shiny side).

1. There should be no other product in the area near the sealing machine.
2. Check that the area/equipment clearance near the machine and all the entries are correct.
3. Tally the product name and batch No. on area/equipment clearance with the label on the bulk container and BPR 1 (batch packaging record).
4. Check that the pack size is as per the master packaging instructions (MPI) requisition.
5. Check whether the operator is wearing a mask, a cap, gloves, and clogs.
6. Confirm that the RH value and temperature recorded by the supervisor is as per the product requirements mentioned in the MPI. If not, inform the area supervisor for corrective action.
7. When all the above requirements are met, countersign the line clearance sheet.

Line Release
After getting the line clearance, the packaging supervisor will start the operation and, after setting the machine and product as per specification, will ask the QAI to check to give the line release.

a. The QAI will confirm that the batch No., expiry date, and the manufacturing date are correct and properly engraved.
b. Check the following randomly:
   1. Pinholes
   2. Empty pockets
   3. Open strips/blisters
c. Perform the leaker test on strips/blisters
d. Check the sealing through the aluminum peel-off test

If all requirements are found to meet the specifications, then the QAI will give the line release by issuing the green release label.

B. Checking for Attribute Characteristics of Sealed Strips/Blisters

1. Audit checks of the strips/blisters during sealing for its attribute characteristics will be performed on an hourly basis.

SOP No. QAS-009.00 Effective date: mm/dd/yyyy
Approved by:

2. Collect a sample of 32 strips/blisters as they are coming out of the machine.
3. The following defects are to be checked:
   a. i. Pinholes
      ii. Empty pockets
      iii. Open strip/blister
   b. Leakers
      - If any defect is found, inform the packaging supervisor and label the lot with "To be sorted" labels and ask for 100% inspection.

      For checking (b), use the following procedure:
      - Perform the leak test on 10 strips/blister as per given procedure, to check that no leakers are present.
      - In case of failure, inform the packaging supervisor and ask to sort, destrip the strips sealed since the last check.
      - Enter the findings on the tablet filling/strip/blistering control chart. Format attached.
      - Leak test to be performed at least twice/shift by the QAI and on an hourly basis by the packaging line boss.
   c. Sealing checks through the aluminum peel-off test
      - After performing the leak test on blisters, allow the machine to run.
      - During the blistering process, collect some samples of blisters as they are coming out of the machine and try to detach the aluminum foil from the middle of the blister to the end.
      - The foil should not peel off easily from the PVC/PVDC.
      - In case it does, inform the packaging supervisor and segregate the portion for sorting and de-blistering.
      - Sealing checks will be performed on an hourly basis.

28.4 ATTACHMENT FORM

QUALITY ASSURANCE IN-PROCESS CONTROL CHART
FOR TABLETS & CAPSULES PACKING

LEVEL 3
SOP No.:

Avg. wt. of three set of 100 tab/cap _______ gm, ______ gm, ______ gm.
Avg. wt. packsize tab/cap. ________________ gm.

Product: ________________ Batch size: ________________ Description of product
Code No.: ________________ Line No.: ________________ Color = ________________
Batch No.: ________________ Line boss: ________________ Shape = ________________
Date started: ________________ Date completed: ________________ Pack size: ________________

Process limits (in case of container pack)
Target wt. = ____________ gm
UCL = ____________ gm
LCL = ____________ gm

		Strip/blistering					Container packing								
							Net wt. of five bottle contents or numbers of fill units								
S. No.	Time	Empty pocket	Empty blister strip	Improper sealing	Leak test	Defective tablet or capsule	1	2	3	4	5	Improper sealing (Caps)	Cracked container	Temp. & RH (area)	Remarks
1															
2															
3															
4															
5															
6															
7															
8															
9															
10															
11															
12															
13															
14															
15															

Description of any corrective action taken = ____

Signature: ________________
QAI

QAS-010

YOUR COMPANY STANDARD OPERATING PROCEDURE

SOP No. QAS-010.00 Effective date: mm/dd/yyyy
Approved by:

TITLE: **Tablets Compression Audit Checks**

AUTHOR: ____________________

Name/Title/Department

____

Signature/Date

CHECKED BY: ____________________

Name/Title/Department

____

Signature/Date

APPROVED BY: ____________________

Name/Title/Department

____

Signature/Date

REVISIONS:

No.	Section	Pages	Initials/Date

SOP No. QAS-010.00 Effective date: mm/dd/yyyy
Approved by:

29.1 PURPOSE

This SOP outlines the procedure for correct in-process audit checks of variables and physical attributes during compression operation of tablets and its requirements.

29.2 RESPONSIBILITY

The area QAI is responsible for carrying out the checks and the QA supervisor is to ensure that the system is followed.

29.3 PROCEDURE

At first the shift QA inspector will check the following.

29.3.1 Area and Equipment Clearance Checks

1. Temperature and humidity
2. Equipment and area clearance
3. Area and equipment release
4. Start-up check to allow the process to run

During compression operation the QA inspector will take samples every hour and confirm the following:

1. Weight variation
2. Hardness
3. Thickness
4. Disintegration
5. Friability
6. Attribute characteristics

29.3.2 Temperature and Humidity Checks

1.1 To check that the temperature and humidity during the operation is within the specified limits.
Tablet: RH—NMT 50% or as per the product specifications; temperature—25°C ± 2°C

1.2 To record the RH and temperature on the control chart for tablet compression. Any deviation observed should be written in the remarks column of the control chart.

29.3.3 Equipment and Area Clearance Checks

2.1 To check and confirm that the equipment and area clearance given by the production supervisor is as per the written procedure.

2.2 To record any deviation observed in the remarks column of the control chart.

29.3.3.1 Procedure

2.3.1 To check the area, equipment, and machine for general cleanliness on every round.

2.3.2 To check the cleanliness of the containers in which granules are kept as well as the containers in which compressed core tablets will be kept.

SOP No. QAS-010.00 Effective date: mm/dd/yyyy
Approved by:

2.3.3 To check proper labeling on the bulk containers in which granules and core tablets are kept.
2.3.4 To check whether the operator is wearing a mask, a pair of gloves, a cap and that the hair is covered properly.
2.3.5 To tally the product name and batch number on the area/equipment/machine clearance with the master manufacturing formula requisition.
2.3.6 To check that the compression room door is properly closed to avoid cross contamination.

29.3.3.2 Area Release

- Perform weight variation on 10 tablets.
- Check physical attributes of tablets.
- Perform disintegration test, hardness, and thickness.
- Also confirm that the correct punches of products are being used by measuring the diameter, and ensuring proper engraving on both sides of the tablet.

If all the above-mentioned tests meet the product specification, then allow the process to run by issuing an area release label and send the samples to the QC for chemical analysis, assay, and dissolution.

29.3.4 Checks during Compression Process

29.3.4.1 Weight Variation Checks

1.1 Initially the machine will be set up by the supervisor as per the written procedure.
1.2 Action limit
  i. Once the machine has been set up and the process is allowed to run, collect the samples from the machine.
  ii. Randomly collect 10 tablets and check their average and individual weights on a balance for checking the correct weight.
  iii. Record the results on control charts.
  iv. Decide the disposition as mentioned below.

Note: When the tablet testing system is used, attach the printed report with the X and R charts and enter the average weight and range on this chart and plot graphically mentioning the container number.

For Group 1: As per the USP limits (NMT 2 tablets ±5% none varies by 10%) tablet weight 325 and above

a. Adjust if:
  i. X crosses ±2% from the theoretical weight.

OR

  ii. 2 tablets cross ±3% from the theoretical weight.

OR

  iii. 1 tablet crosses ±5% from the theoretical weight.
  iv. Range crosses UCL of 50 mg.

For Group 2: As per the USP limits (NMT 2 tablets ±7.5% none ±15%) (tablet weight 130–325 mg/tablet).

SOP No. QAS-010.00 Effective date: mm/dd/yyyy
Approved by:

a. Adjust if:
   i. X crosses ±3% from the theoretical weight.

   OR

   ii. 2 tablets cross ±5% from the theoretical weight.

   OR

   iii. 1 tablet crosses ±10% from the theoretical weight.
   iv. Range crosses UCL of 30 mg.

For Group 3: As per USP limits (NMT 2 tablets ±10% none ±20%) (tablet weight LT 130).

a. Adjust if:
   i. X crosses ±5% from the theoretical weight.

   OR

   ii. 2 tablets cross ±10% from the theoretical weight.

   OR

   iii. 1 tablet crosses ±15% from the theoretical weight.
   iv. The range crosses UCL of 15 mg.
b. Adjust, separate, and check as per laboratory procedure and limits for disposition decision of the compressed tablets since the last check:

For Group 1:

i. If more than 2 tablets cross ±3% from the theoretical weight.

OR

ii. If more than 1 tablet crosses ±5% from the theoretical weight.

OR

iii. If 1 tablet crosses ±10% from the theoretical weight.

For Group 2:

i. If more than 2 tablets cross ±5% from the theoretical weight.

OR

ii. If more than 1 tablet crosses ±10% from the theoretical weight.

For Group 3:

i. If more than 2 tablets cross ±5% from the theoretical weight.

OR

ii. If more than 1 tablet crosses ±10% from the theoretical weight.
   a. Ask to adjust the machine and crush the tablets compressed since the last check.

For Groups 1, 2, and 3:

i. If the laboratory criteria are not met for individual tablets.

29.3.5 Laboratory Procedure and Limits

29.3.5.1 Procedure

If the weight of individual tablets does not meet the action limits, then check the separated portion compressed during that period of time in the following way:

SOP No. QAS-010.00 Effective date: mm/dd/yyyy
Approved by:

i. Collect randomly a sample of ~100 tablets from the separated portion.
ii. Take 20 tablets (out of those hundred tablets) and weigh—calculate the average weight which should be within ±3% of the theoretical weight.
iii. Weigh individually 20 tablets and note down their weights.

Action limits

For Group 1:
Not more than 2 tablets differ from the average weight by more than ±5%. None differs from the average weight by more than ±10%. If the above action limits are not met, then reject and discard that portion of the compressed tablets.

For Group 2:
Not more than 2 tablets differ from the average weight by more than ±7.5%. None differs from the average weight by more than ±15%. If the above action limits are not met, reject and discard that portion of the compressed tablets.

For Group 3:
Not more than 2 tablets differ from the average weight by more than ±10%. None differs from the average weight by more than ±20%. If the above action limits are not met, reject and discard that portion of compressed tablets.

Record any deviation observed in the remarks column of the control chart.

29.3.5.2 Hardness Checks

During tablet compression process, check the hardness of three tablets on the hardness tester or tablet testing system and note down the reading on the control chart.

Action limits

1. None of the three tablets should differ from specification range.
2. If a single tablet differs from the specification range, get the machine adjusted and check the compressed portion by the laboratory procedure.

Laboratory procedure and action limit

1. Collect random samples from the last compressed and separated portion.
2. Record the hardness of 10 tablets.
3. Calculate the average of 10 tablets.
4. Average of 10 tablets should not differ from the specification range.
5. If average hardness of 10 tablets differs from the finished specifications, reject and discard that portion or get the disposition from QA director.

29.3.5.3 Thickness Checks

During tablet compression process, check the thickness of three tablets and note down the reading on a control chart. Note down any deviation observed in the remarks column of the control chart.

Action Limit

None of the three tablets should differ from specifications. If any tablet differs from the specification limits, get the machine adjusted and check the compressed portion by the laboratory procedure.

SOP No. QAS-010.00 Effective date: mm/dd/yyyy
Approved by:

Laboratory procedure and action limits

Collect the samples from the portion compressed in that period and measure the thickness of 20 tablets. Record the mean thickness and range. If the mean thickness differs from the specification limits, reject and discard that portion.

29.3.5.4 Disintegration Test

During the tablet compression process, check the disintegration time twice per shift and note down the reading on a control chart. Note down any deviation observed in the remarks column of the control chart.

Action limits

1. If one or two tablets fail to disintegrate completely within the limit given on the finished product specification, repeat the test on 12 additional tablets. Not less than 16 out of 18 tablets should disintegrate completely within the finished products specification limits.
2. If less than 16 tablets out of 18 tablets disintegrate completely within the given time, then inform the operator to adjust the machine and reject and discard that portion.

29.3.5.5 Friability

a. Check whether the friability apparatus is clean and tidy.
b. Collect two samples for friability A and B each having 20 tablets.
c. Remove any loose dust with the aid of air pressure or a soft brush on the tablet.
d. Weigh samples A and B separately, record the weights.
e. Place them separately in the drums of the friabilator and subject them to 100 rotations (25 rotations/min for 4 min) or as per specification.
f. Remove the tablet.
g. Weigh samples A and B separately.
h. Find out the % loss which should be within the limits as given in the corresponding specifications.
i. Note down the reading on the control chart.
j. During compressor process check friability testing at least twice per shift.

29.3.5.6 Attribute Characteristics

Collect the sample (per sampling plan) ejected from the compression machine and check for attribute characteristics such as:

Critical defect:

a. Foreign matter: Any thing other than the product.

Major defect:

a. Capping: Any tablet showing separation of part of the tablet partially or throughout its horizontal axis. Such tablets may or may not break into two parts.
b. Broken: Any tablet broken into two or more pieces.
c. Picking: Any tablet having depression of 2 mm or more on the surface.
d. Dirty: Any tablet with a spot of a different color (not due to entrapped specks) or oil from the machine.
e. Large specks: Any tablet with entrapped specks or foreign material or 2 mm or more in size.

SOP No. QAS-010.00 Effective date: mm/dd/yyyy
Approved by:

Minor defects:

a. Picking: Any tablets having depression of less than 2 mm in size.
b. Small specks: Any tablet with entrapped specks less than 2 mm in size.
c. Porous surface: Unsmooth surface, resulting in loss of shine.

If any defects are found during the check of 32 tablets, then inform the supervisor to separate and sort. After the corrective action is taken, check as per attached reduced sampling plan.

Action limit

Decide the disposition as per the acceptance plan attached. If rejected, then ask to crush or sort the portion compressed during that period.

Note down the reading on the control chart for tablet compression.

ACCEPTANCE PLAN - A - SINGLE SAMPLING NORMAL INSPECTION ANSI/ASQC Z1.4

			CRITICAL		MAJOR		MINOR5	
SAMPLE SIZE CODE LETTER	LOT SIZE	SAMPLE SIZE	AQL ACC.	0.0% REJ.	AQL ACC.	1.0% REJ.	AQL ACC.	2.5% REJ.
G	3201–10,000	32	0	1	1	2	2	37
H	10,001–35,000	50	0	1	1	2	3	4
I	35,001–150,000	80	0	1	2	3	5	6
J	150,001–500,000	80	0	1	2	6	5	6

Send representative samples to QC for final analysis for releasing bulk products for packaging.

29.4 ATTACHMENT FORM

ABC PHARMACEUTICAL COMPANY

INSPECTION OF PHYSICAL ATTRIBUTES
TABLET/CAPSULES

LEVEL 3
SOP No.: QAS-

No.	Tablet Hardness (kp) Limit:			Tablet Thickness (mm) Limit:			Capsule Lock/Length Limit: (mm)		
1									
2									
3									
4									
5									
6									
7									
8									
9									
10									
11									
12									
13									
14									
15									

Time															
Barrel No.															
Sample size															
Foreign product/odor															
Gross foreign matter															
Unlock capsules															
Thick & thin tablet															
Total critical defects / Sample size	/	/	/	/	/	/	/	/	/	/	/	/	/	/	/
Major															
Capping/laminated															
Sticking															
Dirty (specks/particles)															
Picking >2mm															
Dented capsule															
Pin holes															
Others															
Total major defects / Sample size	/	/	/	/	/	/	/	/	/	/	/	/	/	/	/
Minor															
Chipping															
Sticking															
Picking <2mm															
Small specks															
Porous surface															
Dusty capsule															
Others															
Total minor defects / Sample size	/	/	/	/	/	/	/	/	/	/	/	/	/	/	/

	Friability Limit: %			D.T Limit: min		Temp. RH:		
	Before	After	%	TIME	D.T.	TIME	TEMP °C	RH %
1								
2								
3								
4								
5								

Description of any corrective action:

____

____

____

____

____

ABC PHARMACEUTICAL COMPANY
QA IN-PROCESS X & R CHART

(TABLET & CAPSULES)
LEVEL 3
SOP NO. QAS - 065/127

PRODUCT: ______ PROCESS: ______ SHAPE / MARKING

BATCH NO: ______ CODE NO.: ______ BATCH SIZE: ______ CAPS SIZE: ______

DATE STARTED: ______ COMPLETED ______ MACHINE: ______ CAP: ______

OPERATOR: ______ BODY: ______

PROCESS LIMITS

TARGET WT.: ______ mg

UCL: ______ mg LCL: ______ mg

UCL: ______ mg LCL: ______ mg

X-Chart

Limits	1	2	3	4	5	6	7	8	9	10	11	12	13	14	15
+60mg															
+56mg															
+52mg															
+48mg															
+44mg															
+40mg															
+36mg															
+32mg															
+28mg															
+24mg															
+20mg															
+16mg															
+12mg															
+8mg															
+4mg															
-4mg															
-8mg															
-12mg															
-16mg															
-20mg															
-24mg															
-28mg															
-32mg															
-36mg															
-40mg															
-44mg															
-48mg															
-52mg															
-56mg															
-60mg															

R-chart

Limits		1	2	3	4	5	6	7	8	9	10	11	12	13	14	15
49	52															
45	48															
41	44															
37	40															
33	36															
29	32															
25	28															
21	24															
17	20															
13	16															
09	12															
05	08															
01	04															

	Date/time	Weight of individual tablet/capsule (mg)										X bar	Range	Barrel No.
1														
2														
3														
4														
5														
6														
7														
8														
9														
10														
11														
12														
13														
14														
15														

Weight of 10 empty capsules (mg)						Average	Fill wt.
1		2		3			

Remarks: ______

Corrective action: ______

Average weight= ______ mg

Signature: ______ **QAI**

QAS-011

YOUR COMPANY STANDARD OPERATING PROCEDURE

SOP No. QAS-011.00 Effective date: mm/dd/yyyy
Approved by:

TITLE: **In-Process Checks for Finishing Lines**

AUTHOR: ______________________________

Name/Title/Department

____

Signature/Date

CHECKED BY: ______________________________

Name/Title/Department

____

Signature/Date

APPROVED BY: ______________________________

Name/Title/Department

____

Signature/Date

REVISIONS:

No.	Section	Pages	Initials/Date

SOP No. QAS-011.00 Effective date: mm/dd/yyyy
Approved by:

30.1 PURPOSE

This SOP outlines the procedure for auditing finished products during packaging as well as other checks carried out in the packaging area.

30.2 RESPONSIBILITY

The QAI and the packaging supervisor will be responsible for following the procedure and the QA manager will be responsible for systems compliance.

30.3 PROCEDURE

1. The following checks will be carried out on the packaging lines while carrying out in-process checks in the section.
   - Check that the area is clean and meets all the GMP requirements.
   - Check that all packers/operators are properly dressed and have their hair properly covered.
   - Tally the commodities with master packaging instructions requisition and also confirm that over-printed labels and cartons have been checked by the packaging supervisor prior to packaging.
   - Tally the commodities and product with packaging instructions.
   - Confirm that all the relevant SOPs are being followed.
   - Confirm that the first coded label and cartons have been checked and signed off by the packaging area supervisor.
   - Check that rejected labels, cartons, other commodities, and the product are kept in separate and properly labeled containers.
2. The following procedure is to be followed for audit checks of packed units during the packaging process:
   - Audit checks have to be carried out on each packaging line every hour or on the basis of units packed or processed on an hourly basis as per the attached sample plan (Annexure II).
   - Record all the information on the QA final packaging record (Annexure IV).
   - Collect random samples as per the sampling plan from the packed unit pallet lying in the area.
   - Refer to Annexure I for defects grouping.
   - If any defect is found above the acceptance level, inform the supervisor and quarantine the lot packed between the last two consecutive checks of QAI.
   - Ask for 100% inspection of that portion for sorting out the defectives.
   - Keep the rejected lot segregated with label "To be sorted" until corrective action is taken.
   - After 100% inspection of the quarantined lot/portion, collect the samples and decide the disposition as per the acceptance plan.
   - Enter the defects and action taken on the QA final packaging record of the relevant batch.

SOP No. QCS-011.00 Effective date: mm/dd/yyyy
Approved by:

3. For chemical, microbiological, physical, and reference sample, collect the samples from lines as per SOP for the laboratory.
4. Collect the relevant batch documents from the packaging supervisor after he/she has completed the record, cross check them as per "checklist for batch documents" for finished product (Annexure III), and send the samples along with the batch documents to QC.

30.4 REASON FOR REVISION

First time issued for your company, affiliates, and contract manufacturers.

30.5 ATTACHMENT FORMS

ABC PHARMACEUTICAL COMPANY

SOP No.:
Issued on:
Revision No.:
Initiator name:

ANNEXURE I

DEFECTS (ATTRIBUTES TO BE CHECKED)

From Filling Point

Group I

Clarity
Less count (tablet and capsule)
Less fill check (liquid and granules)
Less removal torque
Chipped brim
Broken plastic cap
Cracked container
Deshaped tubes, dented tubes
Pin holes
Empty unit
Improper cutting
Pitting
Cracking
Edges
Cavity
Hole
Split apex in suppository
Air bubbles

Group II

Liquid spills
P.P. caps broken bridges
Cutmarks on P.P. caps
Improper crimping/sealing
Leak test
Rough surface (suppository)
Concave pit (suppository)
Shiny/dull (suppository)

From Packing

Critical Defects

Wrong batch No./expiry/mfg.
Wrong special stamping
Empty unit
Foreign material
Incorrect number of unit

Major Defects

Wrong commodity
Missing component
Illegible batch No./mfg./expiry
+ Others

Minor Defects

Open label
Crooked label
Torn label
Excessive shrinkage
Smeared overprinting
D-shaped cartons

Excessive Gumming
Double labels
Smeared printing
Improper shrink wrap

ABC PHARMACEUTICAL COMPANY

Page 1 of 2
LEVEL 3
SOP No.: QAS-011
Issued on:
Revision No.:
Initiator name:

ANNEXURE II
MILITARY STANDARD 105-E
GENERAL INSPECTION LEVEL-I
DOUBLE SAMPLING PLAN

Lot Size or Batch Size	Code Letter	Sample Size	Cumul. Sample Size	Critical Defects AQL 0.0%		Major Defects AQL 1.0%		Minor Defects AQL 2.5%	
				ACC.	REJ.	ACC.	REJ.	ACC.	REJ.
2–8	A	2	–	0	1	0	1	0	1
				Single sampling plan					
9–15	A	2	–	0	1	0	1	0	1
				Single sampling plan					
16–25	B	3	–	0	1	0	1	0	1
				Single sampling plan					
26–50	C	5	–	0	1	0	1	0	1
				Single sampling plan					
51–90	C	5	–	0	1	0	1	0	1
				Single sampling plan					
91–150	D	8	–	0	1	0	1	0	1
				Single sampling plan					
151–280	E	13	–	0	1	0	1	1	2
				Single sampling plan					
281–500	F	20	–	0	1	0	1	1	2
				Single sampling plan					

ABC PHARMACEUTICAL COMPANY

LEVEL 3
SOP No.: QAS-011
Issued on:
Revision No:
Initiator name:

Lot Size or Batch Size	Code Letter	Sample Size	Cumul. Sample Size	Critical Defects AQL 0.0%		Major Defects AQL 1.0%		Minor Defects AQL 2.5%	
				ACC.	REJ.	ACC.	REJ.	ACC.	REJ.
501–1200	G	20	20	0	1	0	2	0	3
		20	40	0	1	1	2	3	4
1201–3200	H	32	32	0	1	0	2	1	4
		32	64	0	1	1	2	4	5
3201–10,000	J	50	50	0	1	0	3	2	5
		50	100	0	1	3	4	6	7
10,001–35,000	K	80	80	0	1	1	4	3	7
		80	160	0	1	4	5	8	9
35,001–150,000	L	125	125	0	1	2	5	5	9
		125	250	0	1	6	7	12	13
150,001–500,000	M	200	200	0	1	3	7	7	11
		200	400	0	1	8	9	18	19
500,001 and over	N	315	315	0	1	5	9	11	16
		315	630	0	1	12	13	26	27

Notes:
- ACC. = acceptance number, REJ. = rejection number, Cumul. = cumulative sample.
- If the acceptance number is exceeded on the first sample but the rejection number is not crossed, inspect another sample.
- If the rejection number is exceeded on the first sample, no second sample is taken.

LEVEL 3
SOP No.:
Issued on:
Revision No.:
Initiator Name:

ANNEXURE III
CHECKLIST FOR BATCH DOCUMENTS ALONG WITH SAMPLES (FINISHED PRODUCT)

Product: ____________________ Mfg.: ____________________
Code No.: ____________________ Expiry: ____________________
Batch No.: ____________________ Consignee: ____________________
Stage: Finished Status: Partial/Complete

Packaging Area Documents

1. Batch packaging record number 1 ☐
2. Batch packaging record
   a. Hand packing ☐
   b. Filling and packing ☐
3. Line clearance format ☐
4. First signed coded commodities ☐
5. Reconciliation form (labels/cartons) ☐
6. Packaging report and the release of finished product ☐
7. Packaging material requisition ☐
8. Time sheet ☐

QA In-Process

1. Sampling report (manufacturing side) ☐
2. QA in-process record ☐
   a. QA filling chart ☐
   b. QA final packaging records (for different consignee) ☐
   c. Inspection start-up checklist with release label ☐

Any Other Documents

1. Process change request ☐
2. Memo (sorting or any other) ☐
3. Repacking request/requisition ☐
4. ____________________

Remarks: ____
____
____

Checked and sent by: ____________________ Date: ____________________

Original to QC along samples and documents
cc: QA (In-Process)/Packaging/Production/Stores

ABC PHARMACEUTICAL COMPANY
QA FINAL PACKAGING RECORD

ANNEXURE IV
LEVEL 3
SOP No.:

QC No.: ______ **Line No.:** ______
Product: ______ **Line boss:** ______ **Packaging started:** ______
Code No.: ______ **Consignee:** ______ **Packaging completed:** ______
Batch No.: ______ **Pack size:** ______ **Mfg. date:** ______
No. of units: ______ **Batch size:** ______ **Expiry date:** ______

Commodities used
______
______
______
______

S. No.	CRITICAL DEFECTS / Time / Sample size	1	2	3	4	5	6	7	8	9	10	Description of any corrective action and remarks
1	Wrong commodity											
2	Wrong batch No.:											
	Wrong expiry											
	Wrong manufacturing											
3	Wrong special stamping											
4	Incorrect No. of units in C. Carton											
5	Foreign material											
6	Empty unit											
	Total defects / Sample size											
	MAJOR DEFECTS											
1	Missing component											
	(dosing cup, leaflet, Cannula, etc.)											
2	Illegible Batch No.											
	Illegible mfg. date											
	Illegible exp. date											
3	Others											
	Total defects / Sample size											

	MINOR DEFECTS												
1	Loose or torn labels/carton/box												
2	Double labels												
3	Printing smeared partly												
4	Crooked label												
5	Excessive shrinkage label												
6	Improper shrink wrap												
7	Other												
	Total defects / Sample size												
	Sampled quantity hour											TOTAL	RECEIVED IN QC BY
	Physical/chemical												
	Microbiological												
	Grand total												
	Signature												

QAS-012

YOUR COMPANY VALIDATION STANDARD OPERATING PROCEDURE

SOP No. Val. 100.10 Effective date: mm/dd/yyyy
Approved by:

TITLE: **Internal QA Audit Guideline**

AUTHOR: ______________________

Name/Title/Department

____

Signature/Date

CHECKED BY: ______________________

Name/Title/Department

____

Signature/Date

APPROVED BY: ______________________

Name/Title/Department

____

Signature/Date

REVISIONS:

No.	Section	Pages	Initials/Date

SOP No. Val. 100.10 Effective date: mm/dd/yyyy
Approved by:

31.1 PURPOSE

An internal QA audit system may be defined as a formal program of systematic and comprehensive self-appraisal by a manufacturer, to ensure consistent and reliable production of quality products within his operation. The primary purpose of an internal QA audit program is to assist the management in identifying and correcting deficiencies before they adversely affect the company's products and its reputation for quality.

It is important to realize that an internal QA audit program must be dynamic and will undergo continuing development as sensitivity and awareness to potential problems are increased.

31.2 SCOPE

Purchasing, shipping, registration, maintenance, materials planning, stores, QC, R&D, QA, manufacturing, and packaging.

31.3 RESPONSIBILITY

It is the responsibility of the QA manager to plan and organize an internal QA audit once in a year. The production director, packaging director, maintenance director, R&D director, QC manager, production manager, planning manager, stores manager, shipping manager, purchase manager, and maintenance manager are together responsible to participate or designate in executing a scheduled inspection of the manufacturing, packaging, stores, maintenance, QC, R&D, and general factory premises.

Note: During the SOPs audit, the independence of the auditor from the concerned department has to be maintained.

31.4 PROCEDURE

1. The QA manager will prepare a schedule for audits to cover each department once in a year.
2. During the course of an audit of a section, the officer in-charge or the department head will be present. For audit, the photocopy of the enclosed checklist will be used as a guideline.
3. The audit team will be composed of the following personnels:
    a. QA manager as audit team leader
    b. Team member from production
    c. Team member from packaging
    d. Team member from maintenance
    e. Team member from QC
    f. Team member from PDL
    g. Team member from biotech
4. A report will be issued to the QA director at the end of the audit by the QA manager.
5. This will identify deficiencies and recommend corrective actions to be taken within the specified frame of time.
6. The manager of each department will be presented with a copy of this report. It will then be his responsibility to see that corrective actions are taken. The departmental managers will be responsible for the corrective actions compliance.
7. In presenting any financial justification to the general manager, the departmental managers should refer to the report of the internal QA audit team.

SOP No. Val. 100.10 Effective date: mm/dd/yyyy
Approved by:

8. During the early stage of an audit program, it will seek to define absolute fundamental problems that affect manufacturing and QC operations. As these are resolved, a more critical awareness of potential problems will be highlighted.
9. The tentative audit checklist will be circulated before audit. It is certainly capable of further development and must not be considered as absolute.

31.5 SOP AUDIT

SOP audit should cover all SOPs parameters separately step by step, such as subject, approval, purpose, scope, procedure, and reason for revision.

System SOPs audit is the responsibility of a senior QA officer or designate for all plants except QA system SOPs to maintain the independence of the auditor.

Attachment D should be used as a system audit checklist. In case of nonconformaties, corrective and preventive action report Attachment E will be used. Attachment E will clearly identify the problem (after checking the compliance) mentioning the root cause along with the solution; the solution implementation plan should be written after discussion with the QA manager. The auditor will take signature on Attachment E by the relevant auditee. Follow-up of corrective action implementation will be performed by the SOP auditor on the follow-up date mutually agreed.

31.6 ATTACHMENTS

Attachment A: Checklist for audit
Attachment B: Internal QA audit form
Attachment C: General GMP inspection record
Attachment D: System SOP audit checklist
Attachment E: SOP audit/corrective and preventive action report

31.7 REASON FOR REVISION

First time issued for the ABC Pharmaceutical Industries.

31.8 ATTACHMENT FORMS

Attachment A

SOP No.:
Issued on:
Revision No.:
Initiator name:

CHECKLIST FOR AUDIT

1. Does receiving inspection check incoming shipment to the requirements of the purchase order, specifications, and applicable drawing?
2. Is there a procedure to prevent acceptance of material from unqualified suppliers?
3. Where appropriate, are material weight checks made upon receipt?
4. Do receiving inspection records reflect the reasons for rejection?
5. Are records of receiving activities properly stored?
6. Are inspected items properly segregated from the material awaiting inspection?
7. Is the inspected material adequately identified as to its acceptance or rejection?
8. Is the rejected material adequately controlled?
9. Are the sampling plans employed by quality control sufficient?
10. Is the sampling of received materials performed in a separate area, or with proper controls?
11. Are materials handled, identified, and stored in a manner that will prevent damage, contamination, mix-up, and/or loss.
12. Are accountability records kept to permit traceability?
13. Are stocks reinspected and tested at intervals?
14. Is there a distinct and adequate size label room?
15. Is there label room designed to avoid mix-ups?
16. Is there a sufficient accountability system for labels?
17. Are the material storage and handling procedures documented?
18. Are there specified restrictions on authority for access to labels and other labeling?

AUDIT EXECUTION

1. Evaluate the interpretation of the SOPs with a cross section of people, that is, manager/supervisor/operators.
2. Verify by seeking objective evidence to show compliance to the procedure.
3. Look for evidence of weakness/inconsistencies identified.
4. Evaluate over time the consistent application of procedures, that is, today, last week, and 3 months ago.
5. Verify that communications/services rendered to other departments are properly executed.

PRODUCTION/PACKAGING/IN-PROCESS CONTROL

1. Do the production employees appear to be neat, clean, suitably attired, organized, and operating efficiently?
2. Are there written procedures for production and process control?
3. Are written procedures being followed?
4. Does the production equipment appear to be appropriately designed, constructed, and maintained?
5. Are utensils cleaned to prevent contamination between different product usages?

Attachment A

SOP No.:
Issued on:
Revision No.:
Initiator name:

6. Are appropriate exhaust and vacuum systems employed in operations to minimize air contamination?
7. Are equipment cleaning and use logs being maintained?
8. Are preclearance procedures employed to prevent packaging and labeling mix-ups?
9. Are the batch records kept at the work stations during the entire operation? Are the directions strictly followed?
10. Have time limitations on the holding of processing and in-process items been established and are they being adhered to?
11. Are changes to the issued batch records approved prior to the start of work and documented on the batch records?
12. Are proper gowning techniques and coverings employed?
13. Does the production employed use the acceptable techniques?
14. Are bagged or boxed components stored on the floor?
15. Are lubricants and coolants controlled properly such that they cannot come in contact with product containers, closures, and in-process materials of finished products?
16. Are packaging lines specially or physically separated in a manner designed to prevent mix-ups?
17. Are recorded charts used on time, temperature and pressure monitored processes?

PACKAGING, SHIPPING, AND DISTRIBUTION

1. Do packing and shipping records identify the individuals performing and inspecting the shipping operations?
2. Are adequate storage facilities available and in use to safeguard the quality of the product between final acceptance and shipping?
3. Is a rotated stock system employed to assure that the oldest approved devices are shipped first?
4. Are the packing and shipping containers acceptable to protect the product?
5. Is there a system to assure that only products approved for released are distributed?
6. Do distribution records indicate the location of:
   a. The name and address of the consignee?
   b. The date shipped?
   c. The name and quantity of the item?
   d. The control numbers of the item?

FINAL RELEASE

1. Is there a final inspection performed by quality control on a batch-by-batch basis?
2. Does quality control have written procedures to assure that production records are reviewed?
3. Is there thorough investigation of any discrepancies found during the final review of the production documents?
4. Is the product checked for correct expiration dating?
5. Are records of inspection and data maintained?
6. Is there a classification of defects?

Attachment A

SOP No.:
Issued on:
Revision No.:
Initiator name:

LABORATORIES

1. Are reserve samples of finished products maintained?
2. Have the accuracy, sensitivity, specified and reproducibility of test methods been demonstrated via validation?
3. Are the QC facilities (building) adequate?
4. Do the facilities have adequate equipment for the performance of QC functions?
5. Refer separate SOP for GLP audit.
6. Are test procedures documented?
7. Are raw material file samples maintained for future reference?
8. Are the laboratory reagents and other chemical supplies identified and the expiration dated?
9. Are the laboratory instruments requiring calibration maintained according to the calibration procedures?
10. Is there a written microbial monitoring program for sterile and nonsterile products?

QUALITY ASSURANCE

1. Are the procedures employed by QA documented?
2. Are formal audits of the quality program performed by individuals not having direct responsibility for the program?
3. Are documented corrective actions taken on noted audit deficiencies?
4. Are the procedures for performing audits finalized?
5. Do written procedures governing a document change control system exist?
6. Is there an adequate recall procedure?
7. Is there an adequate stability program?
8. Is batch record data reviewed on an annual basis to determine continual acceptability of quality standards, specifications, manufacturing, and control procedures?

PERSONNEL/TRAINING

1. Is there a training procedure for new employees that includes good manufacturing practices?
2. Is there a refresher training program that includes good manufacturing practices?
3. Are all employees made aware of defects which may occur from improper performance of their respective jobs?
4. Are written job descriptions available?
5. Are personnel in contact with drugs given physical inspections or periodic health checks?
6. Is there a procedure for the exclusion of ill personnel, or personnel who have open lesions, from product contact jobs?

Attachment A

SOP No.:
Issued on:
Revision No.:
Initiator name:

FACILITIES/MAINTENANCE

1. Are there procedures and logs to assure sanitation and pest control?
2. Are adequate washing facilities with hot/cold water, soap, clean, toilets, air driers, or single service towels provided?
3. Is the lighting in all areas adequate?
4. Are adequately constructed waste container located in the appropriate area?
5. Is there a designated area for eating, drinking, and smoking?
6. Is there a formal maintenance schedule of manufacturing equipment? Is it kept visibly near each piece of equipment?
7. Are there positive pressure differentials with all doors leading into less clean areas?
8. Are the positive atmospheric control calibrated and monitored?

VALIDATIONS

1. Have support systems (such as compressed air, vacuum, water) been validated (certified)?
2. Have the environmental control systems been validated (certified)?
3. Have all cleaning procedures been validated?
4. Do all the validation packages include
   a. A detailed description of the item/system (including flow diagrams)?
   b. The intended function of the system?
   c. Installation qualification information?
   d. Operational qualification information?
   e. A preapproved validation protocol?
   f. Cleaning validation protocol?
   g. Explanation of any deviations made to the original signed-off protocol?
   h. Execution of cleaning validation protocols and report?
   i. Validation results?
   j. Quality assurance approval of validation?
5. Is there a formal change control system that addresses changes to validated systems and includes QA review/approval?

Attachment B

SOP No.:
Issued on:
Revision No.:
Initiator name:

INTERNAL QA AUDIT

Team members: ____________________, ____________________.

Dosage forms: ____________________, ____________________.

Products: ____________________, ____________________.

Batch No.: ____________________, ____________________.

Part I: PRODUCT-BASED AUDIT

a. Selection of products from different dosage form, one batch each from

- Tablet
- Capsule
- Cream
- Ointment
- PPS
- Syrup
- Suspension
- Drops
- Suppository
- Injection

Batch selected with the batch No.

____________________ ______________

____________________ ______________

b. Review of (selected) batch documents and related SOPs

- ☐ Completeness
- ☐ Yield calculation
- ☐ Correctness
- ☐ Any nonconformance and their disposition
- ☐ Clarity
- ☐ Any deviations from standards
- ☐ Adequacy
- ☐ Consistency accuracy
- ☐ Batch documents complete with respect to raw materials, packaging commodity, and finishing supplies issues. Manufacturing procedure and all other documents duly signed complete with product and materials reconciliation, and yield calculation. If any deviation, it should be complete with corrective action.

____

____

____

____

____

____

Attachment B

SOP No.:
Issued on:
Revision No.:
Initiator name:

c. Selection of raw materials and packaging material utilized in the batch for checking

Raw materials selected Lot number QC number

Packaging materials selected Lot number QC number

Supplier information (traceable, approved, etc.)

Attachment B

SOP No.:
Issued on:
Revision No.:
Initiator name:

Quality control records

____
____
____
____
____

Analysis records (log books, calculation)

____
____
____
____
____

Receiving report and QC disposition sampling records

____
____
____
____
____

d. Weighing area records of selected batches
Raw materials card checking

____
____
____

Weighing booth log book checking

____
____

Attachment B

SOP No.:
Issued on:
Revision No.:
Initiator name:

Weighing balance calibration record

Weighing area temperature and RH (humidity) records for the weighing of selected batches

Daily calibration record

Area clearance and cleanliness record

Attachment B

SOP No.:
Issued on:
Revision No.:
Initiator name:

Personnel involved and their training record

____

____

____

____

____

e. Manufacturing area records of selected batches

Log book of area and equipment used

____

____

____

Calibration record (balance, pH meter, moisture balance)

____

____

____

____

____

____

Training record of persons involved

____

____

____

____

____

Attachment B

SOP No.:
Issued on:
Revision No.:
Initiator name:

Log books of the machine

____

Punches inspection

____

Sieve inspection

____

f. In-process records of selected batches

Calibration record of instruments

____

Log books entries

____

Attachment B

SOP No.:
Issued on:
Revision No.:
Initiator name:

____

____

____

Record keeping and completeness and correctness of the records in the batch document

____

____

____

____

____

Sampling records

____

____

____

____

____

g. **Packaging records**
Analytical record and disposition

____

____

____

____

____

Approved vendor or not

____

____

____

Attachment B

SOP No.:
Issued on:
Revision No.:
Initiator name:

Packaging line utilized and log books checking for presence and correctness of records related to identified batches

h. QC records

Availability and easy procurement of records from the reference room

- Different batch documents
- Reference samples
- Raw materials analytical/receiving report
- Packaging materials receiving report
- Raw materials reference sample
- Packaging materials vendor approval samples
- Temperature and humidity monitoring of the reference room

Attachment B

SOP No.:
Issued on:
Revision No.:
Initiator name:

i. Stores records

Traceability of selected packaging materials used in identified batches to the supplier

Finished products distribution, record, and reconciliation

Locations control of packaging materials and finished product

Issuance of materials from store (packaging materials card checking) (First in first out (FIFO) maintained or not)

Attachment C

SOP No.:
Issued on:
Revision No.:
Initiator name:

ABC PHARMACEUTICAL COMPANY
INSPECTION RECORD
(GMP)

DEPARTMENT: ______________________________

ROOM NO.: ______________________________

ACTIVITY: ______________________________

PRODUCT: ________________ BATCH NO.: ____________

Walls: ______________________________

Floor: ______________________________

Equipment: ______________________________

Utility lines: ______________________________

Roof: ______________________________

Grills: ______________________________

Exhaust: ______________________________

Ducting: ______________________________

SOPs: ______________________________

Gowning: ______________________________

Other: ______________________________

____

____

Audit performed by:

Signature: ________________ **Date:** ____________

Attachment D

SOP No.:
Issue Date:
Revision No.:
Subject:

SYSTEM PROCEDURE AUDIT CHECK LIST

Following needs update	**YES**	**NO**
Written by designation	☐	☐
Checked by designation	☐	☐
Approved by designation	☐	☐
Purpose	☐	☐
Scope	☐	☐
Responsibility	☐	☐

System procedure audit points	**In compliance**	
	Yes	**No**
1. ______________________	☐	☐
2. ______________________	☐	☐
3. ______________________	☐	☐
4. ______________________	☐	☐
5. ______________________	☐	☐
6. ______________________	☐	☐
7. ______________________	☐	☐
8. ______________________	☐	☐
9. ______________________	☐	☐
10. ______________________	☐	☐
11. ______________________	☐	☐
12. ______________________	☐	☐
13. ______________________	☐	☐

Attachment D

SOP No.:
Issue Date:
Revision No.:
Subject:

14. ______________________________ ☐ ☐
15. ______________________________ ☐ ☐

Attachment as per current practices:

	Yes	**No**
1.	☐	☐
2.	☐	☐
3.	☐	☐
4.	☐	☐
5.	☐	☐

Audited by: ______________

Signature: ______________ **Date:** ______________

Attachment E

SOP No.:
Issued on:
Revision No.:
Initiator name:

QUALITY ASSURANCE

SOP Audit/Corrective and Preventive Action Report

Internal		Audit		Customer complaint		Supplier	

FROM:	TO:
TITLE:	CC:
DATE:	

Subject: ____

____

Revision No.: ____________

1. Statement of problem:
2. Root cause of problem:
3. Solution:

Attachment E

SOP No.:
Issued on:
Revision No.:
Initiator name:

4. Solution implementation action plan:									
5. Monitor dates/actions required by other:									
	Done		**P. Done**		**N. Done**		**N. Required**		
6. Corrective action									
Date:					**Signature:**				

Please attach additional pages if necessary

Auditees signature:

QAS-013

YOUR COMPANY STANDARD OPERATING PROCEDURE

SOP No. QAS-013.00 Effective date: mm/dd/yyyy
Approved by:

TITLE: **Suppository Fill Weight and Physical Attributes Check**

AUTHOR: ____________________

Name/Title/Department

____

Signature/Date

CHECKED BY: ____________________

Name/Title/Department

____

Signature/Date

APPROVED BY: ____________________

Name/Title/Department

____

Signature/Date

REVISIONS:

No.	Section	Pages	Initials/Date

SOP No. QAS-013.00 Effective date: mm/dd/yyyy
Approved by:

Subject: Stability Testing Guideline

32.1 PURPOSE

To assure that marketed products and samples at the customer's end meet label claim, regulatory requirements, and ABC Pharmaceutical Company's standards during stated expiration dating.

32.2 RESPONSIBILITY

It is the responsibility of stability laboratory analysts to follow the procedure. The stability/QC manager is responsible for procedure compliance.

32.3 SCOPE

A. The stability program provides analytical data for active ingredients and other measurable parameters which are affected by time and relevant conditions.
B. Stability studies are performed on
   1. Product to be submitted for registration
   2. Scale-up batches
   3. Product where relevant changes have been made in the manufacturing procedure
   4. Product where significant changes have been made in the packaging material
C. This program also provides data which can support the revision of expiration date, manufacturing excess, container change, etc.

32.4 PROCEDURE

32.4.1 Selection of Batches

1. Stability information for accelerated and long-term testing is to be provided on three batches of the same formulation and dosage form in the containers and closure proposed for marketing.
2. The long-term testing should cover at least 12 months duration at the time of submission.

32.4.2 Test Procedures and Test Criteria

1. The testing should cover those features susceptible to change during storage and that are likely to influence quality, safety, and/or efficacy.
2. All analytical test procedures should be fully validated and the assays should be stability-indicating.

32.4.3 Specifications

1. Limits of acceptance should relate to the release limits (where applicable).
2. The shelf life specifications could allow acceptable and justifiable derivations from the release specifications based on the stability evaluation and the changes observed on storage.

32.4.4 Storage Test Conditions/Frequency

The length of the studies and the storage conditions should be sufficient to cover storage, shipment, and subsequent use (e.g., reconstitution or dilution as recommended in the labeling). See the table

SOP No. QAS-013.00 Effective date: mm/dd/yyyy
Approved by:

below for accelerated and long-term storage conditions and minimum times. Real-time stability study will continue until the proposed shelf life has been covered and the results obtained will be submitted to the registration authority.

Type of Stability Study	Batch Number	Temperature and Relative Humidity	Months				
			Ten	Complete Shelf Life			
Long term	(a) (b) (c)	30°C ± 2°C with (60% ± 5% RH)	1st year 0 3 6 9 12	2nd year 18 24	3rd year 36	4th year 48	5th year 60
Accelerated or short term (I)	(d) (e) (f)	40°C ± 2°C with (75% ± 5% RH)	Three 1 3 6				
(II)	(g) (h) (i)	50°C ± 2°C	Two 1 M 3 M				
For products to be kept refrigerated for long term	(j) (k) (l)	20°C ± 2°C 25°C ± 2°C	Three 1 3 6				

32.4.5 Long-Term Testing

Long-term tests must be carried out on representative batches using the type of packages intended for marketing. A minimum of 12 months long-term stability data will be provided at the time of submission. In this case, the storage is carried out at 30°C ± 2°C/60% ± 5% RH for the entire shelf life.

In this type of study, the observations and testing are done first at zero time then every 3 months during the first year, every 6 months during the second year, and then every year till the end of shelf life.

32.4.6 Short-Term or Accelerated Stability Testing

a. A 6 months stability data at 40°C and/or 3 months data at 50°C will be provided at the time of submission. The details of the storage conditions for accelerated stability studies are
   40°C ± 2°C/75% RH ± 5% RH
   50°C ± 2°C
b. For products to be stored long term under refrigerated condition, the accelerated testing will be conducted at
   25°C ± 2°C
   20°C ± 2°C
c. In accelerated stability, testing is done at zero time, 1 month, 3 months, and 6 months for all required specifications.

SOP No. QAS-013.00 Effective date: mm/dd/yyyy
Approved by:

d. For some dosage forms, especially liquid and semisolid, the study may also need to consider low temperature, that is, freezer (–10°C to –20°C) and refrigerator (2–8°C). However, usually, accelerated studies are less suitable for semisolids and heterogeneous formulations, for example, emulsion.
e. Other storage conditions in stability studies are allowable if justified. The heat-sensitive drug products must be stored under alternative lower-temperature conditions, if justified by the manufacturer, which will eventually become the designated long-term storage temperature. Studies at a low temperature that affects and changes the finished product's properties should be avoided.
f. For products such as solutions and suspensions contained in packs designed to provide a permanent barrier to water loss, specific storage under high relative humidity condition is not necessary.

32.4.7 Production Formula and/or Process Change

Stability samples are required from first product lot following a formula or manufacturing process change, which may have stability significance.

32.4.8 Packaging Materials

The testing must be carried out in the final packaging proposed for marketing. Additional testing of unprotected drug product can form a useful part of the stress testing and pack evaluation, as can studies carried out in other related packaging materials in supporting the definitive pack(s).

32.4.9 Evaluation

A systematic approach will be adopted in the presentation evaluation of the stability information, which should cover, as necessary, physical, chemical, biological, and microbiological quality characteristics, including particular properties of the dosage form (e.g., dissolution rate for oral solid dosage forms).

Significant changes at the accelerated conditions are defined as

a. A 5% potency loss from initial assay value of a batch. All obtained values for all batches should be combined and treated by appropriate statistical tests at 95% confidence limit.
b. Any specific degradation exceeding the product's specification limit.
c. Exceeding the pH limits.
d. Dissolution exceeding the specification limits for 12 capsules or tablets.
e. Failure to meet specifications for appearance and physical properties like
   i Color
   ii. Phase separation
   iii. Resuspendability
   iv. Delivery per actuation
   v. Caking
f. Microbiological specification limit.

32.4.10 Stability Products/Execution

After the product is selected, the stability analyst will initiate the stability protocol and enter the following information:

SOP No. QAS-013.00 Effective date: mm/dd/yyyy
Approved by:

1. Product name
2. Batch number
3. Manufacturing date
4. Expiry date
5. Shelf life
6. Reason for stability study
7. Details of packaging component
8. Testing stations
9. Intervals
    - If any abnormal behavior is observed, it must be discussed and brought to the knowledge of the department's manager and QA director without any delay.
    - Any alteration in the test interval will be done if necessary with the consultation of the QA director.
    - Any abnormality observed which may affect the shelf life will be considered *important* and must be communicated to the department manager and the QA director immediately.
    - Any abnormality which can seriously affect the performance of the product either in the market or which has been planned for marketing should be brought to the knowledge of the managing director by the QA director (or by the department's manager in the absence of the QA director) so that necessary actions can be taken immediately.

1. Color

A product on stability can undergo color changes and may be described as follows:

a. If there is a change in the product's original color to another color, the analyst must describe the color change in detail.
b. If the product maintains the original color but with lighter or darker shade of that color. In this case, there could be three types of changes, that is

Insignilficant:	The change that is not readily noticeable
Significant:	The change that is readily noticeable
Major:	A gross change

The lighter or darker shade can be described as follows:

Fading of color:	This term will be used to indicate the change from the original shade to a lighter shade of the color.
Darkening of color:	This term will be used to indicate the change from the original shade to a darker shade of the color.

2. Odor

The following phrases will be used to describe the change of odor:

- *Characteristic odor:* Represents the original odor of the product
- *Faint characteristic odor:* If the odor is less intense than usual
- *Off odor:* If the odor is different from the characteristic original odor
- Off odor can be subdivided into three types of description depending upon the degree of intensity, insignificant, significant, and severe, to indicate not readily noticeable, readily noticeable, and gross change, respectively.

SOP No. QAS-013.00 Effective date: mm/dd/yyyy
Approved by:

3. Taste

The following phrases will be used to describe the changes in taste:

- *Characteristic taste:* Represents the original taste of the product.
- *Off taste:* If the taste is changed from the original characteristic taste to other taste.
- The term "off taste" will be used with adjectives such as insignificant, significant, and severe to indicate not readily noticeable, readily noticeable, and gross change, respectively.

4. Clarity

Clarity can be described as follows:

- *Clear:* If the liquid contains no undissolved or insoluble matter.
- *Suspended particle:* If there is a separation of some insoluble material but not in the form of precipitates.
- *Precipitates:* If there is a substantial separation of insoluble matter. There can be some sedimentation also.
- *Cloudy precipitate:* If the precipitates are very fine and give the appearance of a cloud. In this case, there can be some sedimentation.
- *Heavy precipitate of fine particles:* Generally no sedimentation is observed.
- *Sedimentation:* When the supernatant liquid is clear but there is some sediment at the bottom. The terms mentioned above will be used with adjectives such as insignificant, significant, and severe, to indicate not readily noticeable, readily noticeable, and gross change, respectively.

5. For the required tests of any product, the stability staff should follow the specifications at the end of the shelf life based on the finished product specifications.

32.4.11 Additional Requirements for the Stability Testing of ANDA Products

The following topics are included:

A. Handling stability failures for products marketed in the United States
   - The stability protocols will be established in coordination with the contracting party and will be followed.
   - Ongoing stability testing will include one batch per year in accordance with the protocols.

B. Trend analysis of stability data
   - The PDL manager will review the batches kept at stability on an ongoing basis as per the protocols. At the end of the year a trend report will be published.

C. Change of supplier(s) of the active pharmaceutical chemical will require the generation of new stability data
   - If the change of the supplier of the active material is unavoidable, the protocols will be prepared in coordination with the contracting party.

D. Registration documents for countries other than the UAE
   - The registration of the products in other countries will be the responsibility of the Julphar registration manager.

E. Fluctuation of the stability data
   - Investigations will be performed to acertain the fluctuation in the stability data of ANDA-related products.

QAS-014

YOUR COMPANY STANDARD OPERATING PROCEDURE

SOP No. QAS-014.00 Effective date: mm/dd/yyyy
Approved by:

TITLE: **Tablets/Capsule Filling In-Process Audit Checks**

AUTHOR: ______________________________

Name/Title/Department

____

Signature/Date

CHECKED BY: ______________________________

Name/Title/Department

____

Signature/Date

APPROVED BY: ______________________________

Name/Title/Department

____

Signature/Date

REVISIONS:

No.	Section	Pages	Initials/Date

SOP No. QAS-014.00 Effective date: mm/dd/yyyy
Approved by:

33.1 PURPOSE

This SOP provides the stepwise procedure for auditing and confirming the tablet/capsule fill checks and physical attributes of containers, closures, and the product and also to check other requirements of the area.

33.2 RESPONSIBILITY

The QA inspector and QA officer will be responsible for following the procedure. The QA manager will be responsible for system compliance.

33.3 PROCEDURE

In-process audit checks will be carried out at the start up and then every hour of the filling process. This procedure is divided into two parts:

A. Following audit checks will be carried out to ensure system and process

A1. Equipment and line clearance checks
   a. The QAI will check the presence of equipment and line clearance issued by the packaging supervisor.
   b. Confirm that the area and equipment are clean.
   c. Confirm the cleanliness and proper labeling on bulk containers in which tablets/capsules are kept.
   d. Confirm "release" label present on the bulk containers.
   e. Check whether operators are wearing masks, gloves, and factory shoes (clogs), and hairs are properly covered.
   f. Tally the product name and batch number on line clearance with master packaging instructions (MPI) requisitions and labels on the bulk container and make sure all entries are correct.
   g. If all requirements are met, then allow the process of filling of tablet/capsule to run.

A2. Humidity and temperature checks
   a. Check the relative humidity (RH)/temperature of the filling area.
   b. RH for the tablet product: NMT 50% or as per product specifications
      RH for the capsule product: NMT 50% or as per product specifications
      Temperature of the area: 25°C ± 2°C
   c. Record the temperature and RH on the control chart (for filling/blistering of capsules and tablet products).

A3. Start up checks for the process to run
When the machine has been set, the packaging supervisor will inform the QAI who will give the release to allow the process to run.

The area QAI (packaging) will collect 10 strips coming out of the machine and will check the strips for correct engraving of batch number and manufacturing and expiry dates. The QAI will then perform the leak test as per SOP: QAS-000. If the leak test results are found within specifications, he/she will issue green label for packaging line release and allow the process to run.

SOP No. QAS-014.00 Effective date: mm/dd/yyyy
Approved by:

In the case of container pack, the QAI will

1. Check the calculation of the tablet/capsule weight for pack size or count the tablets/capsules (as the case may be).
2. Check five filled containers for net weight of tablet/capsule or count the tablets/capsules.
3. Check the correct and proper labeling (if the packaging operation is running side by side with the filling) and if found within specifications will allow the process to run by giving release.

B. Audit checks will be carried out during filling operations

B1. Physical attributes of containers, closures, and the product

a. Confirm that the containers and closures which are used in the packaging are clean and no foreign matter is present inside them.
b. Collect 32 containers randomly from the filled units to check for the defective samples. Check the following defects:
   1. Foreign matter in the containers and closures
   2. Silica gel missing
   3. Chipped or cracked containers or broken plastic cap and improper seal
c. Enter the readings and deviation on in-process control chart for filling of tablets/capsules.
d. If any defect is found, then inform the packaging supervisor and ask to segregate the portion for sorting and corrective action.

Note: The segregated portion could be completely finished units, partially finished, or both.

B2. Less count check for container pack by weight

1. Take three sets of 100 tablets/capsules, and weigh and calculate the average weight of the pack size of tablets/capsules on the basis of the average weight of a single tablet/capsule.
2. Randomly pick up five containers and determine the net weight of the tablet/capsule of each container.
3. Weights should be within ±1% of the target weight calculated.
4. If the weight is not found as per the limits mentioned above, then ask the supervisor to quarantine the portion already filled.
5. Collect samples from quarantine, and decide the disposition of the segregated lots. If they pass, then release the portion.
6. In case of failure, ask to adjust the quantity of tablets/capsules in each quarantined unit.

Note: The segregated portion could be completely finished units, partially finished, or both.

33.4 REASON FOR REVISION

First time issued for the ABC Pharmaceutical Company.

33.5 ATTACHMENT FORMS

ABC PHARMACEUTICAL COMPANY

LEVEL 3
SOP No.: QAS-014
Issued on:
Revision No.:
Initiator name:

ANNEXURE I

DEFECTS (ATTRIBUTES TO BE CHECKED)

From Filling Point
Group I
Clarity
Less count (tablets and capsules)
Less fill check (liquid and granules)
Less removal torque
Chipped brim
Broken plastic cap
Cracked container
Deshaped or dented tubes
Pinholes
Empty unit
Improper cutting
Pitting
Cracking
Edges
Cavity
Hole

Group II
Liquid spills
P.P. caps broken bridges
Cutmarks on P.P. caps
Improper crimping/sealing
Leak test

From Packing
Critical defects
Wrong batch number/expiry/mfg. dates
Empty unit
Foreign material
Incorrect number of units

Major defects
Wrong commodity
Missing component
Illegible batch number/mfg./expiry dates
Others

Minor defects
Open label
Crooked label
Torn label
Excessive shrinkage
Smeared overprinting
D-shaped cartons
Excessive gumming
Double labels
Smeared printing
Improper shrink wrap

QA IN-PROCESS CONTROL CHART
FOR TABLET AND CAPSULE PACKING

LEVEL 3
SOP No.: QAS-014

Average wt. of three set of 100 tab/cap ________ g, ________ g, ________ g
Average wt. pack size tab/cap ________________ g

Product: ____________	**Batch size:** ____________	**Description of product**	**Process limits (in case of container pack)**
Code No.: ____________	**Line No.:** ____________	**Color =** ____________	**Target wt. =** ____________ **g**
Batch No.: ____________	**Line boss:** ____________	**Shape =** ____________	**UCL =** ____________ **g**
Date started: ____________	**Date completed:** ____________	**Pack size:** ____________	**LCL =** ____________ **g**

STRIP/ BLISTERING						DEFECTIVE TABLET OR CAPSULE	**CONTAINER PACKING**							TEMP. & RH (AREA)	REMARKS
S. NO.	TIME	EMPTY POCKET.	EMPTY BLISTER STRIP	IMPROPER SEALING	LEAK TEST		NET WT. OF FIVE BOTTLE CONTENTS NUMBERS OF LIMIT FILLED					IMPROPER SEALING (CAPS)	CRACKED CONTAINER		
							1	2	3	4	5				
1															
2															
3															
4															
5															
6															
7															
8															
9															
10															
11															
12															

QA IN-PROCESS CONTROL CHART
FOR TABLET AND CAPSULE PACKING

LEVEL 3
SOP No.: QAS-014

Average wt. of three set of 100 tab/cap ________ g, ________ g, ________ g
Average wt. pack size tab/cap ______________ g

Product: ____________ **Batch size:** ____________ **Description of product** **Process limits (in case of container pack)**
Code No.: ____________ **Line No.:** ____________ **Color =** ____________ **Target wt. =** ____________ **g**
Batch No.: ____________ **Line boss:** ____________ **Shape =** ____________ **UCL =** ____________ **g**
Date started: ____________ **Date completed:** ____________ **Pack size:** ____________ **LCL =** ____________ **g**

STRIP/ BLISTERING						DEFECTIVE TABLET OR CAPSULE	**CONTAINER PACKING**							TEMP. & RH (AREA)	REMARKS
S. NO.	TIME	EMPTY POCKET	EMPTY BLISTER STRIP	IMPROPER SEALING	LEAK TEST		NET WT. OF FIVE BOTTLE CONTENTS NUMBERS OF LIMIT FILLED					IMPROPER SEALING (CAPS)	CRACKED CONTAINER		
							1	2	3	4	5				
13															
14															
15															

DESCRIPTION OF ANY CORRECTIVE ACTION TAKEN = ____

____

____

____

Signature :____________

QAI

QAS-015

YOUR COMPANY STANDARD OPERATING PROCEDURE

SOP No. QAS-015.00 Effective date: mm/dd/yyyy
Approved by:

TITLE: **Physical Inspection of Attributes of Capsules**

AUTHOR: ______________________________

Name/Title/Department

____

Signature/Date

CHECKED BY: ______________________________

Name/Title/Department

____

Signature/Date

APPROVED BY: ______________________________

Name/Title/Department

____

Signature/Date

REVISIONS:

No.	Section	Pages	Initials/Date

SOP No. QAS-015.00 Effective date: mm/dd/yyyy
Approved by:

34.1 PURPOSE

This SOP provides a procedure for the inspection of physical attributes of capsules.

34.2 RESPONSIBILITY

The QA inspector and the QA officer will be responsible for compliance. The QA manager will be responsible for compliance of this SOP.

34.3 PROCEDURE

After the completion of a process, QA will perform the physical inspection as per the following procedure:

ANSI/ASQC Z1.4 level I (Annexure I) will be used to determine the sample size to be inspected (as per batch size) as well as acceptance and rejection criteria.

a. Inspection of physical attributes to be carried out by the QAI in the humidity-controlled manufacturing area (RH NMT 50% or as per the product's requirement).
b. The QAI should wear gloves and mask during the inspection.
c. Sampled capsules are to be spread over glass paper and under inspection lamp equipped with manufacturing glass. The samples will be observed thoroughly for defects according to the acceptance criteria.

Attribute's Inspection: Capsules

a. Critical defects: AQL 0.0%
    1. Foreign product/odor
    2. Gross foreign matter
    3. Unlock capsules
    4. Cut capsules
    5. Incorrect color/size/shape/print
b. Major defects: AQL 1.0%
    1. Dented capsule
    2. Cracked/powder leakage
    3. Print missing/illegible
    4. Loose closure
    5. Cap/body cutting into one another
c. Minor defects: AQL 2.5%
    1. Dusty capsule
    2. Pinholes (no leakage)
    3. Print smudged

34.4 REASON FOR REVISION

First time issued for the ABC Pharmaceutical Company.

34.5 ATTACHMENT FORM

ANNEXURE I **Level 3**

INSPECTION OF PHYSICAL ATTRIBUTES CAPSULE

Product name: ______________ Batch size: ______________
Code No.: ______________ Sample size: ______________
Batch No.: ______________ Description: ______________

DEFECTS			BATCH SIZE	SAMPLE	ACCEPTANCE	
CRITICAL				**SIZE**	**AQL**	**0.0%**
Foreign product/odor					ACC.	REJ.
Gross foreign matter			35,001 to 150,000	125	0	1
Unlock capsule				125	0	1
Cut capsule			150,001 to 500,000	200	0	1
Incorrect color/size/				200	0	1
Shape/print			500,001 and over	315	0	1
Other defects				315	0	1
Total critical defects / Sample Size						
MAJOR					**AQL**	**1%**
Dented capsule					ACC.	REJ.
Cracked/powder leakage			35,001 to 150,000	125	2	5
Print illegible/missing				125	6	7
Loose closure			150,001 to 500,000	200	3	7
Cap/body cutting into one				200	8	9
Another/pin hole			500,001 and over	315	5	9
Other defects				315	12	13
Total critical defects / Sample Size						
MINOR					**AQL**	**2.5%**
Dusty capsule					ACC.	REJ.
Print smudged			35,001 to 150,000	125	5	9
Other defects				125	12	13
			150,001 to 500,000	200	7	11
				200	18	19
Total critical defects / Sample Size			500,001 and over	315	11	16
				315	26	27
Disposition:			**Conforms** **Does not conform**			

Remarks: ____

____

____

____

____

____

____
Analyst

QAS-016

YOUR COMPANY VALIDATION STANDARD OPERATING PROCEDURE

SOP No. QAS-016.00 Effective date: mm/dd/yyyy
Approved by:

TITLE: **Capsules In-Process Audit Checks**

AUTHOR: ______________________________

Name/Title/Department

____

Signature/Date

CHECKED BY: ______________________________

Name/Title/Department

____

Signature/Date

APPROVED BY: ______________________________

Name/Title/Department

____

Signature/Date

REVISIONS:

No.	Section	Pages	Initials/Date

SOP No. QAS-016.00 Effective date: mm/dd/yyyy
Approved by:

Subject: Introduction to Validation

35.1 PURPOSE

This SOP outlines the procedure for correct in-process audit checks of variable and physical characteristics during encapsulation and its requirements.

35.2 RESPONSIBILITY

The area QAI is responsible for carrying out the checks and the QA manager should ensure that the system is followed.

35.3 PROCEDURE

Area and equipment clearance checks

1. Temperature and humidity
2. Equipment and area clearance

Equipment and area release

1. Start-up check to allow the process to run

During capsule-filling operation

The QAI will take checks every hour and confirm the following:

1. Weight variation
2. Disintegration test (where applicable) twice per shift
3. Lock
4. Length
5. Physical attributes

1. Temperature and humidity checks
   1.1 Check and confirm the RH and temperature during the operation as per the written procedure and the requirement of the area and the product.
   1.2 Record the RH and temperature on the control chart. Any deviation observed should be written in the remarks column of the control chart.

2. Equipment and area clearance checks
   2.1 Check and confirm the equipment and area clearance given by the production supervisor as per the written procedure.
   2.2 Any deviation observed and corrective action taken should be written in the remarks column of the control chart.
   2.3 Procedure
      2.3.1 Check the area, equipments, and machine for general cleanliness.
      2.3.2 Check the cleanliness of the containers in which powders are kept as well as the ones in which capsules will be kept.
      2.3.3 Check proper labeling on the bulk containers in which granules/powders and capsules are kept.

SOP No. QAS-016.00 Effective date: mm/dd/yyyy
Approved by:

2.3.4 Check whether the operator is wearing mask, gloves, and cap, and whether hairs are properly covered.
2.3.5 Tally the product name and batch number on the area/equipment with the master manufacturing formula, requisition, and powder drums.
2.3.6 Check that the capsule-filling cubical door is closed to avoid cross-contamination.

Start-up check to allow the process to run
The QAI will:

- Collect three sets of 10 empty capsules, (representative of the empty capsule to be used in the batch) to calculate the average net weight of the empty capsules. In case of use of a tablet testing system, enter the gross weight of the capsule, that is, the weight of empty shell along with the net weight of powder to be filled in each capsule, when setting up a magazine.
- Calculate the target weight (net weight of powder + empty shell weight) and limits, and will enter it on the control chart for capsule filling.
- Confirm that the machine has been set in the preset limits by the production operator.
- Check the physical attributes of filled capsules (locked, shiny not dusty).
- Check the disintegration time prior to giving release.
- If found within the specification, allow the process to run by giving the area release label.
- Send the assay sample to the QC laboratory from 10–20 capsules.

Checks during capsule filling operation
Weight variation checks

1.1 Initially, the machine will be set by the machine operator as per the written procedure.
1.2 Action limit
  i. Once the machine has been set in its preset limit and the process is allowed to run, collect samples from the machine every hour.
  ii. Randomly collect 10 capsules and check their average and individual weights on the balance for correct weight checks. (In the case of using a tablet testing system for capsule weight variation, attach the printed report along with the control chart.)
  iii. Record the results on the control chart.
  iv. Decide the disposition as follows:

a. Adjust if
  i. X crosses ±3% from the theoretical weight
  OR
  ii. Two capsules cross ±5% from the theoretical weight
  OR
  iii. One capsule crosses ±7.5% from the theoretical weight

b. Adjust, separate, and check as per the laboratory procedure and the limits for disposition decision of the capsule since the last check:
  i. If more than one capsule cross ±7.5% from the theoretical weight
  OR
  ii. If one capsule crosses ±10% from the theoretical weight

SOP No. QAS-016.00 Effective date: mm/dd/yyyy
Approved by:

3. Laboratory procedure and limits
   Procedure
   If the weight of an individual capsule does not meet the action limits, check the separated portion encapsulated during that period of time in the following way:
   i. Collect randomly a sample of approximately 100 capsules from the separated portion.
   ii. Out of these 100 capsules, take 20 capsules and weigh. Calculate the average weight which should be within ±3% of the theoretical weight.
   iii. Weigh individually 20 capsules and note down their weights.

Action limits

1. The individual weight of each capsule is between 90% and 110% of the average weight, that is, ±10% of the average.
2. If not, then individually weigh 20 capsules and keeping their identity, remove the powder content and weigh them again.
   Each individual weight of 20 contents of capsules should be within 90%–110% of the average weight.

Disintegration test

During the encapsulation process, check the disintegration time at start and twice per shift and note down the readings on the control chart. Any deviation observed should be entered in the remarks column of the control chart.

Equipment

Disintegration apparatus with water bath, at 100°C, thermometer.

Procedure

1. Place one capsule each in six tubes of basket and place a disk to each tube with notches upward
2. Hang the basket in a 1000-mL beaker containing deionized water
3. Start the apparatus
4. Continue the test till all the capsules have disintegrated completely
5. Note the time

Action limits

1. If one or two capsules fail to disintegrate completely within the limit given in SOP, repeat the test on 12 additional capsules. Not less than 16 out of 18 capsules should disintegrate completely within the specification limits.
2. If less than 16 capsules out of 18 disintegrate completely within the given time, ask the operator to separate that portion.

Lock checks

The QAI has to check that capsules are being locked properly by the machine by rotating the two shells of a capsule in the opposite direction. In a properly locked capsule, they will not slide. Note down the findings in the control charts.

Length checks

- The QAI has to check whether the length of capsules is as per the specification or not.
- Note down the findings in the control charts for capsule filling.

SOP No. QAS-016.00 Effective date: mm/dd/yyyy
Approved by:

Physical characteristics

The QAI will also check the physical characteristics of capsules on an hourly basis on 32 capsules to ensure that they are not dusty, dented, or ruptured, and that they do not possess any of the defects mentioned in the chart. If so, the inspector will inform the supervisor for corrective action. After the corrective action has been taken, check as per the attached reduced sampling plan.

The representative sample for quality control testing will be taken at the end of the process.

Acceptance Plan C: Single Sampling
Reduced Inspection ANSI/ASQC Z1.4 Level S-4
Acceptable Quality Level +

Sample Size Code Letter	Lot Size	Sample Size	Critical AQL 0.0% ACC.	Critical AQL 0.0% REJ.	Major AQL 1.0% ACC.	Major AQL 1.0% REJ.	Minor AQL 2.5% ACC.	Minor AQL 2.5% REJ.
G	3201–10,000	13	0	1	0	2	1	3
H	10,001–35,000	20	0	1	0	2	1	4
I	35,001–150,000	32	0	1	1	3	2	5
J	150,001–500,000	32	0	1	1	3	2	5

+ = If the acceptance number has been exceeded but the rejection number has not been reached, accept the lot reinstate normal inspection.

Acceptance Plan A: Single Sampling
Normal Inspection ANSI/ASQC Z1.4 Level S-4

Sample Size Code Letter	Lot Size	Sample Size	Critical AQL 0.0% ACC.	Critical AQL 0.0% REJ.	Major AQL 1.0% ACC.	Major AQL 1.0% REJ.	Minor AQL 2.5% ACC.	Minor AQL 2.5% REJ.
G	3201–10,000	32	0	1	1	2	2	3
H	10,001–35,000	50	0	1	1	2	3	4
I	35,001–150,000	80	0	1	2	3	5	6
J	150,001–500,000	80	0	1	2	6	5	5

35.4 REASON FOR REVISION

First time issued for the ABC Pharmaceutical Company.

QAS-017

YOUR COMPANY STANDARD OPERATING PROCEDURE

SOP No. QAS-017.00 Effective date: mm/dd/yyyy
Approved by:

TITLE: **Liquid Products In-Process Audit Checks**

AUTHOR: ______________________________

Name/Title/Department

____

Signature/Date

CHECKED BY: ______________________________

Name/Title/Department

____

Signature/Date

APPROVED BY: ______________________________

Name/Title/Department

____

Signature/Date

REVISIONS:

No.	Section	Pages	Initials/Date

SOP No. QAS-017.00 Effective date: mm/dd/yyyy
Approved by:

36.1 PURPOSE

This SOP outlines the stepwise procedure for the auditing and confirming of the fill volume and physical characteristics of containers, closures, and the liquid products contained within them (syrup, suspension, drops, and solutions).

36.2 RESPONSIBILITY

It is the responsibility of the QA officer and QAI to follow the procedure. The QA manager is responsible for system compliance.

36.3 PROCEDURE

In-process audit checks will be carried out on an hourly basis. This procedure is divided into two portions.

A. The audit checks to be carried out for ensuring the system and process are as follows:
  - A.1 Equipment and area clearance checks will be done
  - A.2 Line release will be done

B. The audit checks to be carried out during filling operations are as follows:
  - B.1 Physical attributes of containers, closures, and the product
  - B.2 Volume checks
  - B.3 Removal torque checks
  - B.4 Leak test

A.1 *Equipment and area clearance checks*

- A.1.1 In the first round of the area, the QAI will check the line clearance granted by the packaging supervisor.
- A.1.2 Check that all packers/operators are properly dressed and their hair are properly covered.
- A.1.3 Check that machine and capper operators are wearing masks and safety goggles.
- A.1.4 Check that only one batch is filled at a time and no other batch or product is lying in the area.
- A.1.5 Check that N_2 gas flushing is done adequately in the bottles (where required).
- A.1.6 Tally the product's name and batch number on line clearance with the MPI requisition.
- A.1.7 Also check for release of the bulk product by QC on the bulk release sheet, its manufacturing and expiry dates.
- A.1.8 If all of the good manufacturing practices (GMP) requirements and area clearance requirements are found to be satisfactory, then allow the process to start, and set up the machine by signing on the line clearance. If any deviation is observed, write in the control chart along with the corrective action taken in the remarks column.

A.2 *Line release*

When the machine is set and the product coming out of it is found to be as per specifications, the packaging supervisor will inform the QAI to grant the line release.

The QAI will check the following:

1. Volume

SOP No. QAS-017.00 Effective date: mm/dd/yyyy
Approved by:

2. Physical attributes of the container and closure
3. Clarity of the liquid
4. Torque
5. Labeling and printing on labels and cartons
6. Leak test (at least once/shift)

If all parameters are within specifications, then the QAI will give the line release and issue a release label to allow the process to run.

During filling operation, checks will be carried out for the physical attributes of filled packs (bottles and caps).

Procedure

1. Check the cleanliness of the primary containers and closure before use.
2. Collect the sample from filled units as per the acceptance plan attached and check defective characteristics of bottles such as broken, chipped brim, cracked, and so on and for caps broken bridges and cut marks on PP caps.
3. Check 13 units hourly for liquid spills, PP caps broken bridges and cut marks on PP caps as per the acceptance sampling plan.
4. Check the clarity of the liquid in five bottles.
5. Enter the findings in the relevant liquid filling control chart.
6. If any deviation is observed, write in the remarks column of the control chart the corrective action taken.

Action limit

1. If in-process control checks cross the cumulative acceptance number in the first check, then ask the operator to adjust the machine and check again.
2. If two successive in-process control fill checks cross the cumulative acceptance number, the units packed since the last check by the line supervisor must be quarantined, resampled, and inspected again and if it still crosses the acceptance number, then quarantine the lot for 100% inspection and inform immediately the packaging supervisor to take corrective action.

Volume checks

Confirm that the machine is running within limits as per the MPIs and in-process product specifications.

Procedure for taking the volume check

1. Collect a specific number of filled bottles corresponding to the number of nozzles of each machine.
2. Determine the volume dispensed in each container by emptying each in a calibrated, dry volumetric cylinder.
3. Record the results on the control chart for liquid filling and decide the disposition as follows:
   a. Adjust if:
      i. X crosses in-process limits.

SOP No. QAS-017.00 Effective date: mm/dd/yyyy
Approved by:

If an individual bottle content is less than that specified in the MPI and in-process control specifications, then ask the operator to adjust the machine and collect eight bottles from the bottles filled since the last check, and check the volume again.

None should be less than the specified limits. If this criterion is not met, then quarantine the portion since the last check and inform the packaging supervisor to take the necessary corrective action.

Note: For pre-set and action limits, refer to packaging in-process control specifications and MPI.

Removal torque checks

During filling operation, check the removal torque of PP capped sealed bottles.

1. Collect five capped bottles from the filling line.
2. Check the torque individually on a standard torque test.
3. Fix the bottle on the torque tester. Hold the cap of the bottle and apply force gently in the anticlockwise direction.
4. Monitor the dial of the torque tester while applying torque.
5. Note the reading when the cap slips on the bottle.

Leak test: Refer to SOP No. ABC-000 for leak test procedures.

Action limit

For the action limit, refer to the acceptance plan attached.

1. If in-process control checks cross the cumulative acceptance limit in the first check, then ask the operator to adjust the machine and resample and check again.
2. If two successive checks for torque cross the cumulative acceptance number, the units packed since the last check by the line supervisor must be quarantined, resampled, and inspected again and if they still cross the acceptance number, then quarantine for 100% inspection and inform immediately the packaging supervisor for the necessary corrective action and note down the findings on the liquid filling control chart.

Standard average removal torque

1. 28 mm PP caps NLT 3 lbs.-in.

36.4 REASON FOR REVISION

First time issued for the ABC Pharmaceutical Company.

36.5 ATTACHMENT FORMS

ABC PHARMACEUTICAL COMPANY

LEVEL 3
SOP No.: QAS-017
Issued on:
Revision No.:
Initiator name:

DEFECTS (ATTRIBUTES TO BE CHECKED)

From filling point
Group I
Clarity
Less count (tablets & capsules)
Less fill check (liquid & granules)
Less removal torque
Chipped brim
Broken plastic cap
Cracked container
Deshaped or dented tubes
Pinholes
Empty unit

Improper cutting
Pitting
Cracking
Edges
Cavity
Hole
Leak test

Group II
Liquid spills
P.P. caps broken bridges
Cutmarks on P.P. caps
Improper crimping/sealing

From packing
Critical defects
Wrong batch number/expiry/mfg. dates
Empty unit
Foreign material
Incorrect number of units
Wrong commodity

Major defects
Missing component
Illegible batch number/mfg./exp. dates
Others

Minor defects
Open label
Crooked label
Torn label
Excessive shrinkage
Smeared overprinting
D-shaped cartons
Excessive gumming
Double labels
Smeared printing
Improper shrink wrap

ABC PHARMACEUTICAL COMPANY

LEVEL 3
SOP No.: QAS-017
Issued on:
Revision No.:
Initiator name:

MILITARY STANDARD SAMPLING PLAN FOR VARIOUS CONTROL LEVELS FOR PROCESS AUDIT FILL CHECKS FROM FILLING POINT

GROUP I

	CHECK		ACCEPTANCE NUMBER	REJECTION NUMBER
Clarity	1st	5 Units	0	2
	2nd	5 Units		
	Combined	10 Units	1	2
Less counts	1st	5 Units	0	2
	2nd	5 Units		
	Combined	10 Units	1	2
Less fill checks (liquid)	1st	As per the number of nozzles of the machines and none should be out of in-process limits for different production dosage forms		
	2nd			
	Combined			
Less fill checks (granules)	1st	3 Units	None should be out of in-process limits	
	2nd			
	Combined			
Less removal torque (Liquid & granules)	1st	5 Units	0	2
	2nd	5 Units		
	Combined	10 Units	1	2
Cotton or silica	1st	5 Units	0	2
	2nd	5 Units		
	Combined	10 Units	1	2
Chipped brim	1st	5 Units	0	2
	2nd	5 Units		
	Combined	10 Units	1	2
Broken plastic cap	1st	20 Units	0	2
Cracked container	2nd	20 Units		
	Combined	40 Units	1	2
Leak test (liquid)	1st	13 Units	0	2
	2nd	13 Units		
	Combined	26 Units	1	2

GROUP II

Liquid spill	1st	13	1	4
PP caps broken bridges	2nd	13		
Cut marks on caps	Combined	26	4	5
Improper sealing/crimping				

QAS-018

YOUR COMPANY STANDARD OPERATING PROCEDURE

SOP No. QAS-018.00 Effective date: mm/dd/yyyy
Approved by:

TITLE: **Document Security and Control and Good Documentation Guidelines**

AUTHOR: ____

Name/Title/Department

____

Signature/Date

CHECKED BY: ____

Name/Title/Department

____

Signature/Date

APPROVED BY: ____

Name/Title/Department

____

Signature/Date

REVISIONS:

No.	Section	Pages	Initials/Date

SOP No. QAS-018.00 Effective date: mm/dd/yyyy
Approved by:

37.1 PURPOSE

The purpose of this SOP is to provide guidelines for SOP and technical documents handling so as to maintain the security and control of the documents.

37.2 RESPONSIBILITY

The departmental managers are responsible to understand and implement the procedure. The QA director is responsible for SOP compliance.

37.3 PROCEDURE

37.3.1 Document Security and Control

1. Effective date system implemented on the first page for SOP and master manufacturing formula (MFM).

37.3.1.1 Documents

Level 1: Policy-related documents are systematically written, checked, and approved. Only authorized personnel can make a photocopy for distribution. All records are maintained appropriately including quality system manual for the company.

Following are the types of documents to be controlled and kept under security. All these documents are kept under security by the documents controlling section.

i.	Manufacturing formula and method	R&D
ii.	Finished production specification	Systems
iii.	Standard test method (STM)	Systems
iv.	Stability report	Systems
v.	Master packaging formula	R&D
vi.	Packaging material specifications	R&D
vii.	Raw material specification (RM)	Systems
viii.	Test method validation report	Systems
ix.	Regulatory affairs documents	Systems
x.	In-process control specification (mfg. & pack.)	Systems
xi.	Raw material purchasing specification	R&D
xii.	Packaging material purchase specification	R&D
xiii.	MFM	R&D
xiv.	Any other document of technical nature considered the property of the company	

The formats can be photocopied or reprinted by the departmental managers and forwarded to the supervisors; however, each format must be signed and dated by the supervisor.

Level 2: SOPs

Note:

- MFM issue for production will be controlled by the QA manager.
- Each planning will send a report to the QA manager.
- After making an entry, the MFM will be issued as a photocopy and the first page will be stamped (red) batch No., date, and signature.

SOP No. QAS-018.00 Effective date: mm/dd/yyyy
Approved by:

Level 3: Quality-related reporting formats stated or not stated in the SOP, but permanently used. The format should be signed by the supervisor and if additional formats are used, mention the reason with signature of the supervisor for additional format.
Level 4: Quality-related reporting formats under trial may or may not be the part of SOP.

Document	Security Check
Level 1	Distribution to be documented and controlled
Level 2	Distribution to be documented and controlled
Level 3	Formats can be photocopied by individual departments for reporting
Level 4	Formats can be photocopied by individual departments for reporting

2. The documents are stored in relevant departments as appropriate.
3. The superseded documents are withdrawn.
4. The changes are incorporated in the documents in the reasons for revision and approved. The following documents identified to have reasons for revision are:
   a. Finished product specification
   b. Raw material specification
   c. Packaging material specification
   d. Master packaging instruction
   e. SOPs
   f. STMs
   g. Stability report
5. For MFM a separate register is maintained for all the changes carried out.
6. The development documents are reviewed according to the written SOP.
7. The files are well maintained and segregated.
8. The responsibilities for the writing, issue, and holding of documents are clearly defined.
9. The indices are maintained to indicate the current status of the documents.
10. The registers are maintained to indicate distribution of records.
11. The documents received from the external organizations are recorded on registers.
12. The technical documents and SOPs produced are formally drafted, reviewed, approved, and authorized for distribution.
13. The documents are issued formally describing their purpose and responsibility as applicable.
14. The documents are reviewed as and when necessary to ensure recipients receive updated copies.
15. All departmental managers are allowed to make one set of photocopy of the document in use to incorporate changes and send it to the QA and R&D department for relevant changes. The changes should be documented in the reasons for revision column.
16. The level 3 and 4 documents can be copied for regular use.

37.4 GOOD DOCUMENTATION GUIDELINE

1. SOPs and the technical documents are to be signed with a blue pen.
2. Overwriting is not allowed in SOPs and other technical documents.
3. As a part of a good documentation skill, if a wrong entry is made, delete the previous entry in a way that it should be visible. Make a new entry beside, below, or above and sign.

SOP No. QAS-018.00 Effective date: mm/dd/yyyy
Approved by:

4. Whitener should not be used to cover mistakes.
5. Changes with a pen over the laminated or nonlaminated SOPs and technical documents are not allowed.

37.5 REASON FOR REVISION

First time issued for the ABC Pharmaceutical Company.

37.6 ATTACHMENT FORM

Page 1 of 1

DOCUMENT CHANGE CONTROL FORM

To : ____________________
From : ____________________
Department : ____________________
Area : ____________________
SOP No. : ____________________

TECHNICAL DOCUMENT:

SUGGESTIONS

____
____
____
____
____
____
____
____
____
____
____
____
____

Signature: ________________________

Date: ________________________

QAS-019

YOUR COMPANY VALIDATION STANDARD OPERATING PROCEDURE

SOP No. QAS-019.00 Effective date: mm/dd/yyyy
Approved by:

TITLE: **Labeling for Incoming Materials, In-Process Products, and Complete Processed Products**

AUTHOR: ________________________________

Name/Title/Department

____

Signature/Date

CHECKED BY: ________________________________

Name/Title/Department

____

Signature/Date

APPROVED BY: ________________________________

Name/Title/Department

____

Signature/Date

REVISIONS:

No.	Section	Pages	Initials/Date

SOP No. QAS-019.00 Effective date: mm/dd/yyyy
Approved by:

38.1 PURPOSE

The purpose of this SOP is to

a. Identify the department responsible for carrying out the labeling at particular steps.
b. Describe the policy and procedure for labeling of incoming materials, quarantined materials, and in-process and finished product.
c. Describe the number and position of the labels on a lot or a batch, especially in case of incoming raw material and packaging materials.

38.2 RESPONSIBILITY

It is the responsibility of production, QC, packaging, materials planning department, maintenance, and QA to carry out the agreed-upon procedure of labeling. The QA manager will be responsible for the SOP compliance.

38.3 GENERAL INSTRUCTIONS FOR PASTING LABELS

38.3.1 Precaution

Never fix labels on the lid or bottom of the container.

38.3.2 Procedure

1. Labels will be printed on the ABC printer using the QA release labeling system
2. Labels printed will be checked against the information of lot in the system
3. All labels will be signed and dated and one label will be kept in a record file for future reference
   a. Peal off the label from the sheet and apply very gently on the surface
   b. Press from one end to the other with a thumb to remove air gaps
   c. Press overturned corners if visible
   d. Fix label on the body of the container only

38.3.3 Labels

The following labels are being used to indicate the status of the item:

1. Production (computer-printed labels): yellow, blue, and white
   a. *Yellow label*: To indicate complete processed bulk product/finished bulk (specimen attached)
   b. *Blue label*: To indicate in-process product (specimen attached)
   c. *White label (printed)*: To indicate rework product (specimen attached)
2. Production (printed labels)
   a. *Clean label (white)*: To indicate the cleanliness of the equipment/area
   b. *Undercleaning label (white)*: To indicate the undercleaning status of the area/equipment
   c. *Production ID*: To indicate the stage of production
   d. *Sterile*: To indicate the cleaning status of the equipment used in the sterile area
3. Store
   a. *Yellow label*: On incoming lot of raw material and packaging materials (receiving computer labels printed)

SOP No. QAS-019.00 Effective date: mm/dd/yyyy
Approved by:

4. Weighing area
   a. White label showing the container of a particular lot, which is "in use," that is, printing "Some quantity of that container has been issued and container is in use, not intact."

38.3.3.1 QC Department

a. *Red label*: Indicating that (*rejected*) materials should not be issued/dispensed/distributed.
b. *Green label (material/product release) computer printed*: For packaging commodities and raw materials (incoming), indicating that these items can be issued, and for bulk product, indicating that it can be taken up for packing.

38.3.3.2 Quality Assurance Department

a. *Orange label*: Indicates the containers from which the sample has been withdrawn, sampled whether it be raw material, packaging material, or bulk product.
b. *QA hold labels*: Yellow label with red printed matter. Materials should not be used until the disposition (GMP deviation observed during inspection) is decided by QA.
c. *"To be sorted" and "to be reworked" labels*: To indicate the disposition of the QC lab or QA for sorting the batch or reworking a portion.
d. *Green label*: Indicating the area/machine or line is released for processing (area/machine/line) after inspecting start-up checks (separate for production and packaging area).
e. *Bulk sampling*: In-process labels for containers with bulk product sample having the information from where the sample has been taken (process) along with product identification.

38.3.3.3 Packaging Department

a. *Semifinished*: Indicating the product is not completely packed and at least one of the packaging component is missing
b. *Pallet label*: Showing the number of pallets packed of the product and the line boss name
c. *Incomplete pack*: Showing the carton on which it is pasted as not has the complete number of finished units
d. *Issued packaging material*: Showing packaging material for a particular line to be supplied after receiving from the store by the packaging department

In addition to these labels, computer-printed labels are used on finished products showing the product's name, batch number, manufacturing date, and expiry date.

38.3.3.4 Maintenance Department

a. *White label with red printing*: Showing that the machine/equipment is under maintenance and not fit for use

SOP No. QAS-019.00 Effective date: mm/dd/yyyy
Approved by:

Responsibility	Procedure
Materials stores	Stores after receiving the packaging or raw materials against the purchase order will paste "Receiving label" (yellow) on each container or bag or whatever the pack size of the incoming material. Each receiving label will have the following details: *Raw material receiving label* a. Material name b. Code No. c. Container No. d. Quantity per container e. ABC Pharmaceutical Company's lot No. f. Manufacturer's name g. Manufacturer's lot No. h. Material receiving date i. Material expiry date j. Supplier name *Packaging material receiving label* a. Code No. b. PO No. c. Supplier name d. Quantity e. Supplier lot No. f. Receiving date g. Container No. h. Lot No.
Quality assurance	The QAI after taking the sample from incoming supplies will paste the "Sampled" label on the container from which the sample has been taken. On approval of the lot, the released label (green) will be pasted on each and every container of raw material by the QA department along with the receiving label, but on bulk supplies (selected items like sucrose, lactose, mannitol, kaolin, maize starch, lactose monohydrate, cetostearyl alcohol, glycerol monostearate, citric acid, cetomacrozol, sodium benzoate, acrosil 200, sorbitol cetyl alcohol) and packaging materials (which are usually packed in bags or those supplies which are issued in bulk to the production or packaging), only three labels will be pasted per pallet. These release labels (printed) will have the following details: *Raw material release label* 1. Material name 2. Code No. 3. QC No. 4. PO No. 5. Lot No. 6. Rechecking date 7. Expiry date 8. Release date 9. Signature *Packaging material release label* 1. Material name 2. Code No. 3. ABC lot No. 4. PO No.

SOP No. QAS-019.00 Effective date: mm/dd/yyyy
Approved by:

(continued)

Responsibility	Procedure
	5. QC No. 6. Release date 7. Signature
	In case the incoming supply is rejected, rejection label will be pasted 100% and the lot will be kept in the rejection area.
Weighing area	At the time of dispensing of released raw material, the dispensing officer will paste the issuance label (white computer-printed or labels filled in by hand) on the portion dispensed with the following details: a. Material name b. Code No. c. QC No. d. Lot No., batch No. e. For which product and batch No. it is being used f. Quantity dispensed g. Tare weight h. Dispensing person's signature i. Date and time of dispensing
	At the time of pasting of these labels, the dispensing officer/issuing personnel will make sure that there is no other label on the dispensed quantity. If present, he will either remove it or deface it before pasting the dispensing labels.
Production	To indicate different stages during manufacturing, the production will use "blue label" for proper identification and status. The status of material will be mentioned on the label against process status, e.g., compressed tablet, blended granules, dried granules, for drying, etc.
	Likewise, on completion, the production department will use a yellow label with the following details: 1. Product name 2. Code No. 3. Process status 4. Batch No. 5. Gross weight 6. Net weight 7. No. of containers 8. Date and signature
	These labels will be pasted 100% on each container of the batch. While taking the sample from the bulk product, the QAI will paste the "Sampled" label on each container from which the sample has been taken out. Production ID (white label with blue printed matter) labels used for the identification of the manufacturing and finishing process. Labels have the following details: Product: Batch no.: Status: Date: Signature:
	Sterile labels are used to identify the cleaning status of the equipment/container and is to be used in injectables. The label has the following information:

SOP No. QAS-019.00 Effective date: mm/dd/yyyy
Approved by:

(continued)

Responsibility	Procedure
	Product: Batch No.: Valid until: Date: Signature:
QA and QC	Upon release of the product by the QC department, green (bulk release label) will be pasted on each container of the bulk product above the yellow label, indicating that this product can be taken up for packaging. The green label will have the following details: 1. Product name 2. Code No. 3. QC No. 4. Batch No. 5. Mfg. date 6. Expiry date 7. Signature of the person responsible for issuing release label 8. Date Bulk release label will be printed by the QC and will be pasted on each container by the packaging area QAI.
Packaging	During packing the packaging department will also use different labels (Attachment I) to show the number of pallet, semifinished product, and incomplete pack carton.
QC	In case a lot (raw material, packaging material, bulk, etc.) fails to meet the specification, the rejection label will be pasted, and the lot will be held in the rejection area till the disposition is decided. In case QC or QA asks for the sorting of the product or for rework of a portion of a product, "To be sorted" labels or "To be reworked" label should be pasted on the lot or portion by the QA (in-process).
QA	During routine inspection, if any major violation of GMP is observed (material/process) which requires that the disposition will be put under QA hold, label will be pasted on portion separated or 100% lot, till the final disposition takes place. The QA hold is removed once the disposition has been decided.

Note: Attachment I refers to the different kinds of labels used by each department, that is, production, packaging, stores, materials planning, quality control, quality assurance, and maintenance.

38.4 REASON FOR REVISION

First time issued for the ABC Pharmaceutical Company.

38.5 ATTACHMENT FORM

ABC PHARMACEUTICAL COMPANY

SOP No.: QAS-019
Issued on:
Revision No.:
Initiator name:

ATTACHMENT I
ISSUANCE LABEL (WEIGHING AREA)
ABC Pharmaceutical Company

Gross wt.:
Tare wt.:
Net wt.:

Material no.:
Code:
QC no.:
Material name:
Product name:
Batch no.:
Weighed by:
Checked by:

Material add by;

Checked by:

RECEIVING LABEL (STORES) RAW MATERIAL
ABC Pharmaceutical Company

Name:
Code no.:
PO no.:
Supplier:
Quantity:
Receiving date:
Manufacturer:
Container No. __________ of __________
Lot No.:

ABC PHARMACEUTICAL COMPANY

SOP No.: QAS-019
Issued on:
Revision No.:
Initiator name:

ATTACHMENT I
RECEIVING LABEL (STORES) PACKAGING MATERIAL

ABC Pharmaceutical Company

Name: Label (Product XYZ)
Code no.:
PO no.:
Supplier:
Quantity:
Receiving date:
Container No. __________ of __________
Lot No.:

"CLEAN" LABEL (PRODUCTION)

ABC Pharmaceutical Company

CLEAN LABEL

Previous product: ____________________ Batch no.: ______________________

Cleaned by: ________________________ Date: __________________________

Checked by: ________________________ Date: __________________________

Validity of cleaning up to: _________________

ABC PHARMACEUTICAL COMPANY

SOP No.: QAS-019
Issued on:
Revision No.:
Initiator name:

ATTACHMENT I
"SAMPLED" LABEL (QA)

ABC Pharmaceutical Company

SAMPLED

Container no.: ________

By/date: ____________

PALLET LABEL (PACKAGING)

ABC Pharmaceutical Company

Product:	
Batch no.:	**Pallet:**
Packaging date:	
Line boss name:	

ABC PHARMACEUTICAL COMPANY

SOP No.: QAS-019
Issued on:
Revision No.:
Initiator name:

ATTACHMENT I
"UNDER MAINTENANCE" LABEL (MAINTENANCE)

ABC Pharmaceutical Company	
UNDER MAINTENANCE	
Date:	Signature:

"CONTAINER IN USE" LABEL (WEIGHING AREA)

ABC Pharmaceutical Company	
CONTAINER IN USE	
Material name:	
Code no.:	Net: _____________ kg
Date:	Tare: _____________ kg
Dispensed by:	Checked by:

ABC PHARMACEUTICAL COMPANY

SOP No.: QAS-019
Issued on:
Revision No.:
Initiator name:

ATTACHMENT I
"ISSUANCE" LABEL (PACKAGING MATERIAL)

ABC Pharmaceutical Company

ATTACHMENT C

Product name:	**Batch no.:**
Packaging material:	**Code no.:**
Quantity:	**Date:**
Signature:	

"TO BE REWORKED" (QA)

ABC Pharmaceutical Company

In-Process Control

TO BE REWORKED

Product:

Code:

Batch no.:

Quantity:

Reason:

Date:

By:

ABC PHARMACEUTICAL COMPANY

SOP No.: QAS-019
Issued on:
Revision No.:
Initiator name:

ATTACHMENT I
"RELEASE" LABEL (RAW MATERIAL)

ABC Pharmaceutical Company

Quality Control Department

RELEASED

Material: ____________________

Code no.: ____________________ Batch/lot no.: ______________

PO no.: ____________________

Receiving date: ______________ Release date: ______________

Signature: ____________________ Expiry date: ______________

"RELEASE" LABEL (PACKAGING MATERIAL)

ABC Pharmaceutical Company

Quality Control Department

RELEASED

Material: C. Carton/48 boxes

Code no.: ____________________ Batch/lot no.: ______________

PO no.: ____________________

Receiving date: ______________ Release date: ______________

Signature: ____________________ Expiry date: ______________

ABC PHARMACEUTICAL COMPANY

SOP No.: QAS-019
Issued on:
Revision No.:
Initiator name:

ATTACHMENT I
"TO BE SORTED" (QA)

ABC Pharmaceutical Company

In-Process Control

TO BE SORTED

Product:

Code:

Batch no.:

Quantity:

Reason:

Date:

By:

"REJECTION" LABEL (QC)

ABC Pharmaceutical Company

Quality Control Department

REJECTED

Material: ______________________

Code no.: ______________________ Batch/lot no.: ______________

PO no.: ______________________

Receiving date: ______________________ Release date: ______________

Signature: ______________________ Expiry date: ______________

ABC PHARMACEUTICAL COMPANY

SOP No.: QAS-019
Issued on:
Revision No.:
Initiator name:

ATTACHMENT I
"BULK RELEASE" LABEL (QC)

ABC Pharmaceutical Company

Quality Control Department

Type: Bulk

RELEASED

Product:

Code no.:

Batch no.:

Batch size:

Container No. __________ of __________

Quantity:

Date:

SEMIFINISHED (PACKAGING)

ABC Pharmaceutical Company

Product name:	**Batch no.:**
Mfg.:	**Exp.:**
Packaging size:	
Date:	
Signature:	

ABC PHARMACEUTICAL COMPANY

SOP No.: QAS-019
Issued on:
Revision No.:
Initiator name:

ATTACHMENT I
INCOMPLETE PACK (PACKAGING)

ABC Pharmaceutical Company

INCOMPLETE PACK

Product name:	**Batch no.:**
Packaging date:	
Contains: ______________________________ **units only**	
Line boss:	**Area supervisor:**

"UNDERCLEANING" LABEL (PRODUCTION)

ABC Pharmaceutical Company

Previous product: Batch no.:

UNDERCLEANING

Date: .... Signature: ....

ABC PHARMACEUTICAL COMPANY

SOP No.: QAS-019
Issued on:
Revision No.:
Initiator name:

ATTACHMENT I
"BULK SAMPLING" LABEL (QA)

ABC Pharmaceutical Company

QUALITY ASSURANCE DEPARTMENT

Product: ______________________________

Code no.: ________________________ Batch no.: ______________________

Batch size/quantity: ______________________

Sampled from ( ) of ( )

Sampled quantity: ______________________

Stage of sampling: ______________________

Time: _________ Date: __________ Signature: ________________________

"IN-PROCESS" LABEL (PRODUCTION)

ABC Pharmaceutical Company

Product name: ___________________________

Code no.: __________________________ Batch no.: ____________________

Process status: ___________________________

Gross wt.: _____________ Tare wt.: _______________ Net wt.: ___________

Container no.: ___________________________

Date: ____________ Signature: _____________________

ABC PHARMACEUTICAL COMPANY

SOP No.: QAS-019
Issued on:
Revision No.:
Initiator name:

ATTACHMENT I
"REWORK" LABEL (PRODUCTION)

ABC Pharmaceutical Company

Product name: ______________________________

Code no.: ______________________________ Batch no.: ______________________

Process status: ______________________________

Gross wt.: ______________ Tare wt.: ________________ Net wt.: ____________

Container no.: ______________________________

Date: ______________ Signature: ________________________

"QA HOLD" LABEL

ABC Pharmaceutical Company

QUALITY ASSURANCE

QA HOLD

Product: Batch no.:

Date: By:

Reason

ABC PHARMACEUTICAL COMPANY

SOP No.: QAS-019
Issued on:
Revision No.:
Initiator name:

ATTACHMENT I
PRODUCTION DEPARTMENT
FINISHED BULK LABEL

ABC Pharmaceutical Company

Product name: ______________________________

Code no.: ____________________________ Batch no.: ______________________

Process status: ______________________________

Gross wt.: ______________ Tare wt.: ________________ Net wt.: ____________

Container no.: _________of__________________

Date: _____________ Signature: _______________________

"AREA/MACHINE RELEASE" (QA) PRODUCTION SIDE

ABC Pharmaceutical Company

Quality Control Department

RELEASE

MANUFACTURING

Product name: _____________________

Code no.: _________________________ Batch/lot no.: ________________

Area: ____________________________ Machine: ____________________

Process: __________________________ Date: ________________________

Signature: _________________________

ABC PHARMACEUTICAL COMPANY

SOP No.: QAS-019
Issued on:
Revision No.:
Initiator name:

ATTACHMENT I
"MACHINE/LINE RELEASE" (QA) PACKAGING SIDE

ABC Pharmaceutical Company

QUALITY ASSURANCE

RELEASE

PACKAGING

Product name: ____________________

Code no.: ____________________ Batch no.: ____________________

Line no.: ____________________ Machine: ____________________

Mfg. date: ____________________ Expiry date: ____________________

Date: ____________________ Signature: ____________________

"STERILE"

ABC Pharmaceutical Company

Product name: ____________________ Batch no.: ____________________

STERILE

Valid until:

Date: Signature:

QAS-020

YOUR COMPANY STANDARD OPERATING PROCEDURE

SOP No. QAS-020.00 Effective date: mm/dd/yyyy
Approved by:

TITLE: **Preinspection Audit Guidelines for ANDA Products**

AUTHOR: ____________________________________

Name/Title/Department

____

Signature/Date

CHECKED BY: ____________________________________

Name/Title/Department

____

Signature/Date

APPROVED BY: ____________________________________

Name/Title/Department

____

Signature/Date

REVISIONS:

No.	Section	Pages	Initials/Date

SOP No. QAS-020.00 Effective date: mm/dd/yyyy
Approved by:

39.1 PURPOSE

To provide a guideline to perform an audit to ensure that abbreviated submissions to the FDA are adequately supported and documented. Submissions subjected to this procedure include New Drug Application, Abbreviated New Drug Applications and Supplements.

39.2 RESPONSIBILITY

All departmental managers are responsible for following the procedure.

39.3 INSTRUCTIONS

Before a preinspection audit, department managers will make sure that all necessary documents as per the procedure are approved, current, and are available at the point of use and in the master file.

39.4 PROCEDURE

Check that the following are as per the given requirements.

39.4.1 Technical Documents

Check that the following approved and current documents are available at the point of use and in the master document file.

a. Raw material specifications with the test method (active and inactive)
b. Finished product specifications
c. Standard test methods
d. Packaging commodity specifications
e. Master manufacturing formula and method
f. Master packaging instructions
g. Validation master plans
h. Cleaning validation master plan

39.4.2 Raw and Packaging Materials

a. The raw and packaging materials receiving log entry for each component will be checked to ensure that
   - Sufficient quantity was received to justify the usage in the batch.
   - The date received is consistent with the date of usage in the batch.
   - The vendor of the material is the source indicated in the submission.
   - Material meets all the requirements given in the SOP.
   - Check that the material receiving report was prepared as per the SOP.
   - Check that raw material was sampled and stored as per the procedure.
b. The raw material inventory card for each component will be checked to ensure that
   - The amount entered on the card is consistent with the usage reflected on the batch record.
c. The following documentation will be collected for review.
   - For each raw material (active or inactive) used in the batch:
     - The raw material/packaging material receiving log entry.

SOP No. QAS-020.00 Effective date: mm/dd/yyyy
Approved by:

- The raw material/packaging material inventory card.
- The QC test report and the manufacturer's certificate of analysis.

39.4.3 Product Batch Record

a. The batch record(s) for the batch(es) supporting the submission will be examined to ensure that
   - The batch document audit checklist is available.
   - The approved master manufacturing formula is available.
   - All the released raw materials issued by stores and supporting documents are available and correct.
   - All manufacturing steps were performed according to the formula.
   - Each step "Performed by" and "Checked by" initials with date are entered.
   - All calculations are correct.
   - All weighing printouts are available.
   - All the supporting documents, for example, sterilization/lyophilization/drying/CIP and SIP, and so on, printouts are available and correct.
   - All in-process tests were performed as per procedure, and records are available.
   - Equipment cleaning logs for each piece of equipment used in the manufacture of the batch are available.
   - All test reports (in-process, finished product, etc.) with raw data are also required.
   - Approved master packaging instructions are available.
   - All the released packaging materials issued by stores and supporting documents are available and correct.
   - Reconciliation record for labels/cartons/product.
   - All calculations are correct.
   - All calculated yields are within established limits and, if not so, are adequately investigated and explained on the batch record.

b. The manufacturing procedure used for manufacturing the test batch is the same as the master formula to be included in the submission.
c. An accountability test for the batch will be performed, including the bulk and packaged product, for comparison with the actual yield entered on the batch record.

39.4.4 QC Record

The QC test report, certificate of analysis and test specifications will be checked for each component to ensure that

a. The raw material (active and in-active) was released by QC before usage in the batch.
b. If the date of initial QC testing exceeds the date of usage by >1 year, it must be verified that the appropriate raw material (RM) retesting has been performed.
c. It must be verified that each required test on the test specification has been performed.
d. The manufacturer's lot number should be checked on each of the following documents to ensure consistency:
   - The RM receiving log entry
   - The QC report
   - The certificate of analysis

SOP No. QAS-020.00 Effective date: mm/dd/yyyy
Approved by:

e. Check analyst register, logs, register.
f. All test reports (in-process, finished product, raw material, etc.) will be examined to ensure that
   - All raw data including calculations, curves, chromatograms, and so on. Adequate raw data exist to support all test results.
   - The physical description entered in the reports is consistent with the actual description of the product.
   - All the required testing was performed, and is within established specifications.
g. Environmental monitoring records.

39.4.5 Stability Batch Record

Stability test results will be examined to ensure that
a. Stability testing protocol is available.
b. All other documents as described in Section 39.4.4.
c. Adequate raw data exist to support all test results.
d. The physical description entered in the reports is consistent with the actual description of the product.
e. All the required testing was performed, and is within established specifications.
f. All stability reports, with raw data (calculations, curves, chromatograms, etc.).
g. Stability test specifications are also required.
h. The stability samples will be examined, and the following information recorded:
   - Physical description of the product (color, shape, scoring, markings, etc.).
   - Packaging system for sizes on stability (bottle manufacturer, bottle size, closure type, presence of desiccants, etc.).
i. Concluded stability test report.

39.4.6 Validation Record

The following validation records should be reviewed:
a. Test method validation reports
b. In case of injectable
   - Bioburden method validation on raw material and finished product
   - Endotoxin method validation on raw material and finished product
   - Sterility method validation using the actual product
c. IQ/OQ and PQ validation record
d. Facility qualification
e. Heating ventilation air conditioner (HVAC) qualification
f. Utilities/water qualification
g. Process validation record
h. Cleaning validation record
i. In case of injectable
   - Heat distribution and heat penetration records for mobile vessel
   - Heat distribution and heat penetration records for autoclave (each load)
   - Heat distribution and heat penetration records for hot sterilization tunnel (each size of ampoule/vial)
   - Heat distribution record during SIP cycle for freeze dryer

SOP No. QAS-020.00 Effective date: mm/dd/yyyy
Approved by:

- Temperature uniformity validation records for freeze dryer
- Line qualification by media fill

j. Cleaning validation

39.4.7 Other Documents

a. If the submission includes a bioavailability study, the packing slip or letter which accompanied the sample(s) submitted to the bioavailability testing laboratory will be copied.
b. For submissions with bioavailability studies, the following must be reviewed:
   - The lot number on the packing slip or letter which accompanied the sample(s) submitted to the bioavailability testing laboratory should be identical to the lot number entered in the bioavailability report.
   - The date the sample(s) were sent to the bioavailability testing laboratory should be consistent with the manufacturing and testing dates for the batch.
   - If the report includes a physical description of the product, it should be consistent with the actual description.
c. The actual packaging components used in the stability study should be consistent with that described in the submission, and those entered on the stability report.
d. Trends for viable count, nonviable count, surface monitoring, and personnel monitoring.
e. Sterilization information and data.
f. Approved drawings for facility, water plant and lines, HVAC, men/material movement, and so on.

39.5 REASON FOR REVISION

First time issued for the ABC Pharmaceutical Company.

QAS-021

YOUR COMPANY STANDARD OPERATING PROCEDURE

SOP No. QAS-021.00 Effective date: mm/dd/yyyy
Approved by:

TITLE: **Operational Qualification Guideline**

AUTHOR: ______________________________

Name/Title/Department

____

Signature/Date

CHECKED BY: ______________________________

Name/Title/Department

____

Signature/Date

APPROVED BY: ______________________________

Name/Title/Department

____

Signature/Date

REVISIONS:

No.	Section	Pages	Initials/Date

SOP No. QAS-021.00 Effective date: mm/dd/yyyy
Approved by:

40.1 PURPOSE

The present document describes the guidelines that will be applied to the operational qualification of pharmaceutical utilities and equipments.

40.2 RESPONSIBILITY

It is the responsibility of ABC Pharmaceutical Company's plant managers, QA manager, engineers, and validation team members (QA, QC, production, packaging, stores, maintenance, and water treatment plant) to follow the procedure as a guideline.

The QA officer will draft the initial protocol with cGMP requirements embedded in it. The protocols will be sent to the concerned engineers/members as appropriate for technical checks. The amended protocols will then be sent back to the QA officer for finalization. The finalized protocols will be signed off by the reviewers. The protocol will be authorized by the QA director.

The approved protocol will be executed under the supervision of plant managers/engineers/QA managers and the QA officer.

40.3 PROCEDURE

Some optional sections could be added, for example, a paragraph describing the test and measure equipments used.

The content of the operational qualification is detailed hereunder.

40.3.1 Document Description

This describes the operational qualification documents. This description includes

- Responsibility
- A paragraph stating the objective of the document (description of subject)
- A paragraph specifying the scope of the document, that is, the exhaustive list of the concerned equipment(s), or categories of equipments (qualification scope)
- Any other observation
- Written by
- Verified by

40.3.2 Operational Checks

This description includes

- Supplier-fixed set points (if required)
- Alarms and safety checks (if required)
- Analog transmission inputs (if required)
- Machine setting checks as per user manual
- Critical operating parameters
- Actual test runs of machine
- Any other check
- Conclusion

SOP No. QAS-021.00 Effective date: mm/dd/yyyy
Approved by:

- Any deviation
- Comments
- Performed by/verified by

40.3.3 Standard Operating Procedures

The following drafted procedures should be available at the time of operational qualification:

- Operation
- Cleaning
- Maintenance
- Calibration (if required)
- Training
- Other SOP (if applicable)
- Conclusion
- Any deviation
- Comments
- Performed by/verified by

40.3.4 Training

Identify the names of the staff for training on the following at the time of operational qualification:

- Operation
- Cleaning
- Maintenance
- Calibration (if required)
- Other (if applicable)
- Conclusion
- Any deviation
- Comments
- Performed by/verified by

40.3.5 Software/Hardware (Automated Process Controls)

- Software/hardware structural testing (optional)
- Software/hardware validation reports/certificate
- Manufacturer audit reports
- Factory acceptance criteria reports
- Conclusion
- Any deviation
- Comments
- Performed by/verified by

The above checks may be amended as applicable or other checks may be added. All documents shall be verified.

SOP No. QAS-021.00 Effective date: mm/dd/yyyy
Approved by:

The equipment is already in place mostly since 1980; therefore, the retrospective data are collected where possible.

Note: Refer to the attached Format I (Utilities and Equipment Qualification Requirements Matrix).

40.4 REASON FOR REVISION

First time issued for the ABC Pharmaceutical Company.

40.5 ATTACHMENT FORM

ABC PHARMACEUTICAL COMPANY

Level 3
SOP No.: QAS-
Issued on:
Revision No.:
Initiator name:

FORMAT I

QUALIFICATION REQUIREMENT MATRIX (FOR UTILITY)

Each utility within the scope of this document must be assigned a qualification level based on the use and application of the utility. Qualification and requirements details are as follows:

	Level 1	Level 2	Level 3
Criteria	The following utilities are classified as level 1: • Water for injection • Purified water • Clean steam • Other, as designated by the QA manager	The following utilities are classified as level 2: • Product contact compressed gas system • Bulk solvent (product contact/ingredient) distribution systems • Water pre-treatment systems • Autoclave/retort cooling water for parenterals • Nutritional ingredient water and culinary steam	A utility that requires qualification as defined by the QA manager. Examples may include • Selected waste removal/processing systems • Selected vacuum systems • Selected nonproduct contact utilities
Initial requirements	• IQ/OQ/PQ • Establishment of system on extended monitoring program (for pharmaceutical water systems)	• IQ/OQ/PQ	• IQ • Other, as determined by QA manager

QUALIFICATION REQUIREMENT MATRIX (FOR EQUIPMENT)

	Level 1	Level 2	Level 3
Criteria	Equipment where an undetected change in equipment performance poses a significant risk to the product	Equipment where a change that may be detected in equipment performance poses a significant risk to the product	Packaging equipment
Initial requirements	• IQ/OQ/PQ	• IQ/OQ • Other (PQ) by QA manager	• IQ • Other (OQ, PQ), as determined by QA manager

QAS-022

YOUR COMPANY STANDARD OPERATING PROCEDURE

SOP No. QAS-022.00 Effective date: mm/dd/yyyy
Approved by:

TITLE: **The Retention of Records and Documents**

AUTHOR: ______________________________

Name/Title/Department

____

Signature/Date

CHECKED BY: ______________________________

Name/Title/Department

____

Signature/Date

APPROVED BY: ______________________________

Name/Title/Department

____

Signature/Date

REVISIONS:

No.	Section	Pages	Initials/Date

SOP No. QAS-022.00 Effective date: mm/dd/yyyy
Approved by:

41.1 PURPOSE

The purpose of this SOP is to ensure that records and documents are retained for an appropriate and specific period of time. The procedure applies to SOPs, technical documents, and documented records of QA and QC departments.

41.2 RESPONSIBILITY

All concerned departmental managers are responsible for following the procedure. The QA director is responsible for SOP compliance.

41.3 PROCEDURE

1. All batch records must be retained for a minimum of *shelf life of the batch +2 years* from the corresponding date of manufacture of the batches. The records that are maintained in batch files include
   a. All manufacturing records
   b. All packaging records
   c. All in-process control records (line clearance, in-process checks)
   d. All QC records (analytical reports, release certificates, labels, cartons, etc.)
   e. All weighing area records
2. Any complaint related to the batch will be filed with the QA manager with the conclusion and will be maintained in the "Complaint File" and will not be destroyed.
3. The master product specifications and the master test method (original) will be maintained by the manager documentation. If there is any change in the product specifications and the test method, then the old original copies will be filed in "*STM and product specification*." These records/files will be maintained for a minimum period of 7 years from the *date of revision*.
4. Validation reports (of equipment, process, systems, etc.) will be maintained in separate files and these files will be updated whenever there is a revalidation maintaining the previous records. The records will be maintained by the validation manager for an indefinite period.
5. Test method validation reports will be filed with the method validation group and will be kept for an indefinite period.
6. Stability records of a product will also be maintained for an indefinite period in the R&D laboratory.
7. The destruction of SOPs (master and copies) will be indicated by an asterisk in a column of returned SOPs. All the changes in the SOPs are maintained under "Reasons for Revision" in the current SOP.
8. All technical packages will be retained with systems for an indefinite period. The manager documentation will keep the records for the superseded documents.
9. Product development records shall be retained with the R&D laboratory director for an indefinite period.
10. All the packaging material approval records (development) will be maintained by the R&D manager and will be retained for an indefinite period.
11. All the vendor approval samples records will be maintained by the QC laboratory in "Archive" and will be maintained for an indefinite period.
12. Prescribing and patient information (a sample of each version) or patient guidance leaflet will be retained for a minimum of 7 years from the date it ceases to be authorized for use.

SOP No. QAS-022.00 Effective date: mm/dd/yyyy
Approved by:

They must be accompanied by an indication of the dates between which they were authorized for use (drug information department).

13. Adverse drug event records (correspondence, records, and reports) or authentic copy must be retained indefinitely with the Regulatory Affairs and Drug Information Departments. No such records may be destroyed without the authority of the director QA or the general manager of the ABC Pharmaceutical Company.
14. Calibration and maintenance records of equipment in the QC laboratory, QA, and R&D laboratory will be maintained by the manager calibration for a period of 2 years, after which it will be destroyed.
15. Purchase requisitions will be kept by the purchase department for a period of 2 years after the date of delivery, after which it will be destroyed.
16. The ABC Pharmaceutical Company Intranet Documentation System request form will be kept for an indefinite period after submitting to the manager documentation.

Tabular Presentation of Quality Records

S. No.	Document Type	Responsibility	Duration
1.	Batch records and analytical reports	QC	Shelf life +2 years
2.	Validation records	Validation	Indefinite
3.	Product specification (Standard control procedure (SCP)) and STMs	QC/R&D	For 7 years after revision
4.	Prescribing and patient information	Drug information department	For 7 years
5.	Adverse drug event records	Regulatory affairs/ drug information	Indefinite
6.	Product development record (product file) (design control)	R&D	Indefinite
7.	Calibration record of the equipment	Maintenance and calibration	2 years
8.	Test method validation reports	R&D	Indefinite
9.	Stability records	R&D	Indefinite
10.	Product package/technical	Regulatory affairs	Indefinite
11.	Purchase requisition	QA/QC/R&D	6 months after receipt of requested item
12.	Technical package	RM	Indefinite
13.	Temperature and humidity records of file sample room	QC	1 year
14.	Management review	QA	Indefinite
15.	Handling of nonconformities	QC	Shelf life +1 year
16.	Customer complaints	QA	Indefinite
17.	Purchasing requisition	Purchase	2 years
18.	Approved vendors list	QC	Indefinite
19.	Temperature monitoring records (batch file)	QC	Shelf life +1 year
20.	Training records	QA	Indefinite
21.	Servicing (trouble shooting)	R&D	Indefinite
22.	Internal audits	QA	Indefinite
23.	Transmission request form	Website	Indefinite
24.	New drug application (NDA) and abbreviated new drug application (ANDA) files submission	QA	Indefinite

SOP No. QAS-022.00 Effective date: mm/dd/yyyy
Approved by:

41.4 DOCUMENTATION

The documents will be destroyed by shredding when the duration of retention is complete.

41.5 REASON FOR REVISION

First time issued for the ABC Pharmaceutical Company.

QAS-023

YOUR COMPANY STANDARD OPERATING PROCEDURE

SOP No. QAS-023.00 Effective date: mm/dd/yyyy
Approved by:

TITLE: **Good Manufacturing Practices in the Premises**

AUTHOR: ______________________________

Name/Title/Department

____

Signature/Date

CHECKED BY: ______________________________

Name/Title/Department

____

Signature/Date

APPROVED BY: ______________________________

Name/Title/Department

____

Signature/Date

REVISIONS:

No.	Section	Pages	Initials/Date

SOP No. QAS-023.00 Effective date: mm/dd/yyyy
Approved by:

42.1 PURPOSE

This SOP outlines the checklist of good manufacturing practices (GMP) in the premises to be followed as a guideline.

42.2 RESPONSIBILITY

All employees working in the plant are responsible for following the procedure. All departmental managers are responsible for implementing the procedure.

42.3 PROCEDURE

42.3.1 Frequency

All checks will be done randomly. Note the findings on the GMP deviation form and pass it to the related person for corrective actions.

42.3.2 General

Check the following:

42.3.2.1 Audit the Checklist

1. Only those workers (coming into direct contact with the product) who have been medically examined and cleared are allowed to work in the area.
2. Workers in the area must be wearing clean uniforms and caps.
3. All areas are cleaned as per respective cleaning procedures.
4. No eating, smoking, or unhygienic practices are allowed in the manufacturing area.
5. The facility as well as the lockers provided in the facility must be clean. SOP of cleaning is followed.
6. All manufacturing operations are carried out in separate areas intended for the purposes.
7. All apparatus and equipment to be used in the operation have been cleaned as per their respective SOPs.
8. The contents of all vessels and containers used in the manufacturing and storage between manufacturing stages are identified by clearly legible labels having all the required information.
9. All machines and equipment bear clear legible clearance label, having all the required information.
10. All respective SOPs are followed.
11. Insecticutors on every entrance are available as appropriate.
12. The under construction or maintenance areas must be well segregated from the manufacturing area.
13. All the specific areas have the required temperature, relative humidity (RH), and air pressures.
14. Clogs are in use instead of street shoes.
15. Gauges/measuring equipment have valid calibration tags on them.
16. Dispensing and all the processing are being carried out in the prescribed and dedicated areas.
17. Line clearance procedure must be strictly followed and counter checked to avoid mix-up.

SOP No. QAS-023.00 Effective date: mm/dd/yyyy
Approved by:

18. Devices for the presence of labels and overprinting were challenged prior to the initiation of work.
19. Effective segregation of printed materials from not printed and rejected ones.
20. Utilities lines are clearly marked and labeled.
21. Storage of product (bulk and finished) is under their labeled or prescribed storage conditions.
22. All records must be completed at the time of action.
23. Employees have undergone training in the following areas:
    a. GMPs
    b. SOPs
    c. Sterile area techniques and respective SOPs are followed.
24. Employees have full knowledge of their job functions.
25. Each department is well maintained for cleanliness and is spacious enough for equipment and operations.
26. The area and equipment are clean at the end of the day's work.
27. Specific procedures for the cleaning of major equipments are followed.
28. Daily calibration records of balances are up to date.
29. Correction to writing errors are made by crossing out, with initial, date, and reason (if necessary).
30. All the products have status label.
31. All the log books (equipment/area usage records) are updated.

42.3.2.2 Stores and Weighing Area

1. Weighing and measuring equipments used are of appropriate accuracy and are calibrated.
2. All incoming supplies of raw material and packaging commodities are stored as per SOP.
3. Stores receiving labels are fixed on all containers and sampled labels are affixed as appropriate.
4. Release labels are fixed on approved items and rejected labels on rejected items.
5. All rejected items are kept segregated from other materials.
6. Cleaning in the area is done according to schedule and is very effective.
7. Workers (coming into direct contact with the product) are wearing clean masks.
8. The door of the weighing booth during dispensing is closed to avoid cross contamination.
9. Dispensed materials are kept in locked, clean cages with proper label.
10. Log books of the weighing area and calibration records of weighing balances are updated.
11. Raw material card entries are on time, complete and correct.
12. Temperature and RH are within the requirements and limits and the records are maintained.
13. Storage of the raw materials, packaging materials, bulk products, and finished products is done appropriately.

42.3.2.3 Solid Dosage Production Area

1. Return ducts or exhaust in the production and filling areas are working properly.
2. RH and temperature are within limits, as per requirements of the product.
3. Workers (coming into direct contact with the product) are wearing clean masks.
4. Operators wear disposable gloves while touching the product.
5. All the products have status labels.
6. All the other respective SOPs are being followed.

7. All the log books (equipment/area usage records) are updated.
8. Batch-related documentation is on time, complete, and correct.

42.3.2.4 Liquid Dosage Production Area

1. Temperature indicators having valid calibration tags.
2. Workers (coming into direct contact with products) are wearing clean masks and gloves.
3. All the products have status label.
4. Batch-related documentation is on time, complete, and correct.
5. All the log books (equipment/area usage records) are updated.

42.3.2.5 Packaging

1. All packed units after completion of batch are kept in the finished goods area in stores.
2. All coded and uncoded labels and boxes are kept in cages with identification label.
3. Workers coming into direct contact with the product are wearing masks.
4. Operators wear disposable gloves while handling the product.
5. There is proper physical separation/segregation between the lines.
6. Temperature and humidity records of dry and soft packaging areas are updated and maintained.

42.3.2.6 Cephalosporin and Penicillin Production Area

1. Operators are wearing a mask and gloves and using a beard cover when necessary.
2. Relevant cleaning procedure for the area and equipment should be followed.
3. Work areas are clearly labeled.
4. All vessels and utensils are labeled as to their cleanliness status.
5. Manufacturing instructions are at hand during processes and are approved and accurately followed.
6. Vessels should be labeled with product lot number and the stage of processing.
7. Air inlets and outlets are functional.
8. Material holding containers are intact and can be closed completely.
9. Materials and personnel flow is dedicated and being followed.
10. Temperature and humidity are within the limits.

42.4 DOCUMENTATION

42.4.1 Good Documentation Skills

1. The correct ink color and type and the proper writing instruments are used (pencils must not be used).
2. Additional documents (e.g., charts, printout) are properly included.
3. The printing and writing can be easily read.
4. Errors are properly corrected with one line through the original entry, initialed, dated, and the reason explained (if necessary).
5. Calculations are reviewed and verified.
6. Numbers are properly rounded off; the correct significant figures are used.
7. The spelling of the product names and other words is correct.
8. Lot numbers and product ID codes are double checked and are correct.
9. Information is recorded as it is obtained.

SOP No. QAS-023.00 Effective date: mm/dd/yyyy
Approved by:

10. Proper formats for date and time are used.
11. All abbreviations are approved and standardized.
12. Blank spaces are properly handled.
13. The original entry is made into the official record.

Good documentation skills should produce proper records with the following characteristics:

a. Permanent
b. Accurate
c. Prompt
d. Clear
e. Consistent
f. Complete
g. Direct
h. Truthful

42.5 STERILE PRODUCTION AREA

1. In clean room operations, employees should be correctly attired according to the relevant SOP.
2. A fresh set of garments is used on each entry into the clean room.
3. A set of garments ready for use should be sealed and labeled.
4. The sterile production area should be in a good state of repairs and neat.
5. A person ill or suffering from cold or cough or having an open or bandaged wound should inform to the supervisor and should not be allowed in the clean room areas.
6. The relevant cleaning procedure and schedule for the area and equipment should be followed.
7. Cleaning and sanitization agents should be as per the SOP and labeled with the expiry date.
8. Records to be maintained for the preparation of cleaning and sanitization agents.
9. No simultaneous opening of doors occurs on the clean and less clean class.
10. All the alarm systems are functional.
11. Work areas are clearly labeled.
12. Product and product components are exposed only where protected by laminar air flow (LAF) stream providing an air quality of 100 or better.
13. Handling and all the working practices should be such as to avoid contamination and generation of particles.
14. All vessels and utensils are labeled as to their cleanliness status.
15. Manufacturing instructions are at hand during processes and are approved and accurately followed.
16. Vessels should be labeled with the product lot number and the stage of processing.
17. Relevant sterilization and other charts and printouts are fully labeled, verified, and approved.
18. Valid calibration labels on all equipment.
19. Air inlets and outlets are functional.
20. Material-holding containers are intact and can be closed completely.
21. Materials and personnel flow is dedicated and being followed.

SOP No. QAS-023.00 Effective date: mm/dd/yyyy
Approved by:

22. Air differentials are maintained between the different areas of processing as per their levels of cleanliness.
23. Temperature and humidity are within the limits.
24. Dispensing and all the processing are being carried out in the prescribed and dedicated areas.
25. Line clearance procedure should be strictly followed and counter checked to avoid mix-up.
26. Devices for the presence of labels and overprinting challenged prior to initiation of work.
27. Effective segregation of printed materials from nonprinted and rejected materials.
28. Utility lines are clearly marked and labeled.
29. Storage of product (bulk and finished) is under their labeled or prescribed storage conditions.
30. All records must be completed at the time of action.
31. Watches, bracelets, jewelry, or rings should not be worn during work.

42.5.1 "Do"s to Be Followed

1. Do wear identification card all the time at work.
2. Do follow clothing instructions before entering the plant.
3. Do wash hands before entering and leaving the personal facility.
4. Do read, understand and follow standard operating procedures related to your work.
5. Do follow GMP at work.
6. Do observe personal hygiene at work.
7. Do keep your nails clean and cut them every week.
8. Do get a haircut every month.
9. Do keep your beard trimmed every day before coming to work.

42.5.2 "Do Not"s to Be Followed

1. Do not take food and drinks inside the plant; in particular, do not eat chewing gum.
2. Do not enter into the plant in personal clothing, street shoes, and unwashed hands.
3. Do not go to the bathroom and outside the plant in your working uniform, head cover, and clogs.
4. Do not enter the plant if suffering with some contagious disease or a lesion over the skin.
5. Do not carry an antidust face mask, disposable hand gloves, and other working tools to your lockers.
6. Do not keep medicines inside the locker, except those prescribed for personal use by the doctor.
7. Do not carry batch record and support documents in your pockets and lockers.
8. Do not remove medicines from the line for personal use.
9. Do not taste or touch raw materials for personal experience during dispensing and processing.
10. Do not change labels from the container unless authorized.
11. Do not remove items from the pallets unless authorized.
12. Do not cough inside the plant without handkerchief.
13. Do not use cosmetics and jewelry at work.
14. Do not keep used sanitary towels inside the lockers.

SOP No. QAS-023.00 Effective date: mm/dd/yyyy
Approved by:

15. Do not sign batch records unless authorized.
16. Do not use sampling and other tools dedicated for the potent drugs production area and vice versa.

42.6 REASON FOR REVISION

First time issued for the ABC Pharmaceutical Company.

QAS-024

YOUR COMPANY VALIDATION STANDARD OPERATING PROCEDURE

SOP No. QAS-024.00 Effective date: mm/dd/yyyy
Approved by:

TITLE: **Fill Volume, Color, and Particulate Matter in Drops Bottles**

AUTHOR: ______________________________

Name/Title/Department

____

Signature/Date

CHECKED BY: ______________________________

Name/Title/Department

____

Signature/Date

APPROVED BY: ______________________________

Name/Title/Department

____

Signature/Date

REVISIONS:

No.	Section	Pages	Initials/Date

SOP No. QAS-024.00 Effective date: mm/dd/yyyy
Approved by:

Subject: Fill Volume, Color, and Particulate Matter in Drops Bottles

43.1 PURPOSE

This SOP outlines the stepwise procedure to check and confirm the fill volume, color, and particulate matter in drops bottles.

43.2 RESPONSIBILITY

The area QAI and the QA officer will be responsible for ensuring correct fill checks, color, and checks for particulate matter of the product and other related checks. The QA manager is responsible for system compliance.

Equipment Required

1. Volume checks
   a. 400 mL beakers
   b. Acetone
   c. Deionized water
   d. Six 25 mL calibrated cylinders
   e. Six 50 mL calibrated cylinders
2. Particulate matter/color check
   a. Optical checking hood

Frequency

Every hour of filling of drops.

43.3 PROCEDURE

After line clearance granted by the packaging supervisor, the QAI/QA officer will check and confirm the following steps during the first round in the area to confirm the line clearance.

1. The area and equipment are to be kept clean.
2. Commodities mentioned in the master packaging instructions requisition tallies and no other commodity should lie in the area except the ones mentioned in the requisition.
3. Filtration set up, gas connections (where required) and their flow is satisfactory.
4. Operators and workers in the area are properly dressed and wear masks, gloves, and goggles, if necessary, taking all precautionary measures.
5. Equipment and area sanitization/cleaning record is satisfactory.

When all the above requirements are found satisfactory then sign on the line clearance form.

43.3.1 Fill Volume Check

1. Check the volume by emptying the contents (depends upon number of nozzles) directly into a calibrated cylinder.
2. If the average volume and range crosses the in-process limits, then ask the concerned to adjust the machine and inform the packaging supervisor.

SOP No. QAS-024.00 Effective date: mm/dd/yyyy
Approved by:

TABLE 43.1

Pack Size (mL)	Preset Limit (mL)	In-Process Limit (mL)
10	10.3–10.5	10.0–10.5
15	15.0–16.0	14.0–16.0
15 (susp.)	15.0–17.0	15.0–16.0
30 (drops)	30.0–31.0	30.0–32.0

3. If the volume of individual bottle or average crosses the laboratory limits ask the concerned operator to adjust the machine and separate the portion filled before. Inform the packaging supervisor and manager.
4. Enter the findings on a control chart for liquid filling.
5. Also confirm the readings taken by the line boss as per the written procedure. If any deviation is observed, then enter in the remarks column of the control chart.

Note: For fill checks limits and range limits refer to Table 43.1.

43.3.2 Color and Particulate Matter Checking

After conducting the volume variation check, carry out the color and particulate matter check on the same sample.

Stopper the measuring cylinder and check the solution in optical checking hood divided into two portions, one should have a white background and the other black. The illumination should be from a shielded 100 W bulb day light/quality light source, at a point 25.4 cm (10 in.) from its source.

Inspect the solution with a shaking motion invert the measuring cylinder and retain it to the upright position under an optical checking hood for 10–12 s in front of a white background and then in black background to inspect the following:

- Fiber
- Particles (black or white)
- Any other foreign matter

Note any fine flocculent material in the solution that tends to float, hang, or slowly settle downwards. Check the clarity of the solution, there should be no haze, the solution must be sparkling clear. The color should be as specified. Enter your findings in the control chart for liquid filling.

43.4 REASON FOR REVISION

First time issued for the ABC Pharmaceutical Company.

43.5 ATTACHMENT FORMS

ABC PHARMACEUTICAL COMPANY

LEVEL 3
(Liquid Filling)
SOP No.: QAS-

QA IN-PROCESS CHART

Product: ________________________ **Pack size:** ____________ **Target Vol:** ____________mL
Code No.: ____________ **Batch No.:** ____________ **Batch size:** ____________ **Ucl:** ____________ mL
Date started: ____________ **Date completed** ____________ **Line No.:** ____________ **Lcl:** ____________ mL

1	Date/Time	Volume in Individual Bottle (mL)								Av. Volume (mL)	Range
2											
3											
4											
5											
6											
7											
8											
9											
10											
11											
12											
13											
14											
15											

Description of any corrective action taken: ________________________________

____

____

____

____

Average Volume = [mL] **Signature** **QAI**

QA001/ZA

LEVEL 3
SOP No.: QAS-

QA IN-PROCESS CONTROL CHART

LIQUID FILLING

	Date/ Time	Clarity	Removal Torque	Average Removal Torque	Chipped BRIM	Cracked Container	Liquid Spills	Presence of Gas	Cut Mark on PP Caps	Area Temp. °C	Leak Test	Sign	Remarks
1													
2													
3													
4													
5													
6													
7													
8													
9													
10													
11													
12													
13													
14													
15													

Description of any corrective action: ____________________

____

____

____

____

Signature **QAI**

QAS-025

YOUR COMPANY STANDARD OPERATING PROCEDURE

SOP No. QAS-025.00 Effective date: mm/dd/yyyy
Approved by:

TITLE: **Random Sampling**

AUTHOR: ______________________________

Name/Title/Department

____

Signature/Date

CHECKED BY: ______________________________

Name/Title/Department

____

Signature/Date

APPROVED BY: ______________________________

Name/Title/Department

____

Signature/Date

REVISIONS:

No.	Section	Pages	Initials/Date

SOP No. QAS-025.00 Effective date: mm/dd/yyyy
Approved by:

Subject: Random Sampling

44.1 PURPOSE

The purpose of this SOP is to describe the measures taken by the ABC Pharmaceutical Company management to use random sampling plan and statistical technique for sampling of raw and packaging material, and in-process and finished product.

44.2 RESPONSIBILITY

The QAI is responsible for following the procedure and the QA manager is responsible for the SOP compliance.

44.3 PROCEDURE

To ensure that the sample selected from a lot is representative of that particular lot, it is essential to select it in a manner such that each part has an equal chance of being selected regardless of its quality or ease or difficulty of obtaining it.

1. If the lot is in one container: Select parts proportionally from all areas of the container. Take sample from the top, middle, bottom, and sides in approximately equal proportions.
2. If the lot is in several containers: Refer to the table given below to determine the minimum numbers of containers from which the samples are to be selected.

Total Number of Containers	Number of Containers from Which Sample Should Be Taken
1–5	All
6–10	5
11–50	7
51–100	10
101–250	15
Over 250	20

3. The table will show the number of containers to be opened.
4. Assign number to all the units (containers) in a lot to uniquely identify each unit.

Tables of random numbers are useful in obtaining random samples when containers in the lot are identified separately.

44.3.1 Instructions for Using Table of Random Numbers

1. Containers in a lot must be numbered or placed in rows so that numerical positions can be established.
2. Select any point in the table of random numbers. (This can be accomplished by putting a pencil at random on a digit in the table.)

SOP No. QAS-025.00 Effective date: mm/dd/yyyy
Approved by:

3. Once this starting point is picked, movement in any direction will give a random sample of numbers.
   Example:
   a. A lot of 45 containers numbered 01 to 45. A random sample of 7 is desired.
   b. Suppose the point selected is number 15 (Attachment I (page 1 of 2), 2nd column, 9th row).
   c. If we move down, the samples selected would be numbers 19, 01, 24, 31, 12, 42. (We could also move up or across.)
4. When a number encountered in the table of random numbers is greater than the total number of containers in the lot, this number should be discarded and the next number taken.
5. When a number encountered is a duplication of a number already chosen, the duplicate should be discarded and the next number taken.

44.4 ATTACHMENTS

Attachment I: Table of random numbers. Lot size up to 1–100 containers (two pages)
Attachment II: Table of random numbers. Lot size up to 101–1000 containers (two pages)

44.5 REASON FOR REVISION

First time issued for the ABC Pharmaceutical Company.

44.6 ATTACHMENT FORMS

Attachment I

Table of Random Numbers (Lot size up to 1 to 100 containers)

83	78	46	81	90	31	79	26	70
19	9	10	55	11	23	75	59	82
9	94	27	85	63	83	58	96	88
30	91	89	64	46	97	81	10	19
62	20	1	82	76	85	51	23	88
36	42	83	12	64	28	51	42	42
19	9	48	17	10	41	93	52	14
9	42	42	36	70	19	8	30	16
45	15	27	76	25	31	65	24	89
87	78	7	52	25	62	42	16	66
31	74	34	53	23	28	35	70	99
42	19	100	71	41	95	98	47	84
79	1	73	47	46	89	60	69	64
27	85	20	43	20	76	69	78	1
83	56	53	53	18	2	65	3	27
75	65	73	89	89	90	61	25	8
93	24	72	72	46	48	31	85	50
62	31	54	79	67	25	60	78	86
17	61	90	25	97	14	71	70	37
25	12	43	37	85	17	80	76	30
87	54	55	4	53	20	62	5	91
22	42	5	95	10	64	44	57	27
62	43	15	62	97	52	94	75	91
4	28	82	91	59	92	5	32	85
72	18	14	85	95	44	20	14	15
31	17	65	55	14	72	74	79	26
24	15	97	81	99	52	80	76	51
36	4	76	71	88	25	61	1	59
79	38	97	39	91	9	98	96	96
60	37	82	29	18	26	44	62	66
48	92	3	17	28	17	91	93	8
21	2	17	87	81	9	70	31	5
61	30	62	72	69	8	22	6	74
56	31	46	69	43	33	1	89	20
88	18	30	17	63	94	35	84	37
13	74	14	84	49	84	47	59	87
54	55	48	82	40	23	88	10	100
85	97	18	47	10	7	24	19	86
65	44	92	49	81	7	60	82	31
80	51	9	32	38	81	19	17	10
9	60	86	25	7	85	6	56	15
80	58	73	42	20	89	94	63	39
65	50	71	95	99	75	57	6	27
20	67	57	32	72	71	66	25	69
51	16	69	20	100	14	59	80	70
73	3	43	93	88	28	7	35	48
32	38	63	37	74	73	65	80	94
67	1	61	57	89	33	36	23	6
34	45	8	77	83	79	99	57	34

(continued)

Table of Random Numbers (Lot size up to 1 to 100 containers)

76	91	6	42	24	31	97	69	58
85	75	1	17	42	68	90	74	79
33	93	43	78	4	7	26	5	11
12	25	87	98	91	52	9	37	63
9	81	25	65	35	50	7	80	83
22	72	73	7	1	54	70	19	42
7	86	64	62	18	80	17	47	70
33	34	49	30	45	34	91	70	11
31	60	87	60	85	38	87	9	34
90	7	79	71	58	31	87	56	55
86	24	61	40	19	31	72	44	64
87	32	71	16	81	55	2	32	19
59	61	78	1	69	4	26	49	51
31	88	50	47	56	70	6	98	43
21	62	45	16	70	5	98	26	48
80	69	64	46	31	16	45	58	64
79	95	7	92	14	90	6	53	27
57	59	51	40	47	12	18	48	4
78	2	9	93	86	36	52	43	98
74	18	2	94	4	57	5	56	15
22	8	96	84	65	40	3	29	29
40	98	77	74	81	6	44	36	57
82	59	49	27	90	91	7	13	55
96	78	43	11	83	47	21	68	24
30	90	65	3	62	93	94	23	72
61	11	51	89	34	41	68	14	24
19	48	59	94	62	87	1	30	6
69	25	65	8	87	21	12	28	94
1	50	48	21	68	5	48	52	72
39	66	86	33	77	87	100	43	18
30	15	49	47	74	14	81	67	20
89	43	42	31	57	80	15	18	2
31	60	93	57	21	8	85	54	79
46	41	16	82	22	45	89	42	62
100	35	47	82	25	7	30	65	67
88	68	83	98	35	20	82	51	46
43	3	12	47	91	12	56	41	81
2	78	11	76	11	96	82	9	79
8	42	84	4	94	69	70	33	90
37	17	27	52	62	39	82	22	1
54	55	76	68	85	76	68	74	93
53	76	78	37	94	89	94	12	90
8	69	78	89	99	80	32	92	35
90	71	25	69	92	20	48	51	36
6	98	64	6	8	28	79	33	99
80	6	54	97	52	68	92	79	41
79	55	88	52	90	69	48	53	97
24	97	3	31	46	92	83	87	73
42	42	32	5	33	35	61	63	74

Attachment II

Table of Random Numbers (Lot size up to 101 to 1000 containers)

358	480	856	356	851	362	261	467	293
913	306	817	555	751	164	675	375	908
507	687	494	577	17	456	246	428	936
136	20	33	329	445	642	963	721	541
324	500	396	66	107	347	282	347	993
885	744	770	913	610	599	873	100	842
131	157	356	269	28	328	775	154	502
419	180	1	284	302	980	555	182	81
458	886	97	859	446	414	176	854	311
344	546	52	400	315	669	336	608	940
91	359	193	137	464	810	463	131	129
336	317	10	764	100	271	902	286	441
279	24	651	627	775	315	884	264	831
751	652	580	340	685	643	198	221	230
205	199	345	924	745	756	789	552	770
513	101	416	264	136	713	146	798	778
790	325	743	332	985	305	77	745	893
778	644	885	102	472	94	823	749	382
545	290	515	716	412	536	927	20	563
939	750	891	222	141	445	288	312	535
569	101	88	663	276	253	681	7	4
17	198	542	581	739	196	143	737	980
379	778	7	987	541	588	930	963	454
463	157	954	200	848	735	282	657	684
756	732	790	864	503	258	761	866	204
866	861	94	979	634	548	609	615	259
628	526	463	92	994	720	693	902	293
747	452	1	598	326	358	973	847	567
910	814	85	213	829	555	941	729	305
500	976	918	831	352	834	411	577	893
760	197	998	555	27	23	978	59	506
312	273	262	979	857	588	569	966	140
145	644	440	82	442	727	344	476	122
271	259	736	385	85	822	515	634	35
504	177	360	915	673	475	911	52	811
401	486	906	931	86	106	189	151	571
577	187	538	589	232	898	20	662	66
103	345	545	53	470	574	532	556	487
764	741	251	6	19	8	438	367	737
501	244	768	950	151	542	868	894	605
370	902	840	90	338	798	253	896	173
929	468	214	748	607	510	465	444	917
234	392	863	795	832	823	908	391	170
38	366	247	894	531	372	967	378	455
931	71	290	20	508	817	597	655	296
216	808	161	613	952	343	906	541	746
102	422	112	709	933	36	746	194	486
73	142	8	437	252	960	961	649	567
547	400	582	536	991	864	697	753	832

(continued)

Table of Random Numbers (Lot size up to 101 to 1000 containers)

928	254	693	681	360	123	386	873	100
913	436	313	799	853	392	279	289	860
544	898	226	270	403	128	682	367	732
33	611	199	609	628	505	649	764	120
196	821	12	59	580	211	983	723	965
750	317	759	125	629	944	605	608	825
719	633	70!	842	992	561	804	265	203
112	732	997	390	191	624	216	788	431
575	606	679	128	970	485	727	751	632
779	23	822	302	399	39	418	526	884
58	735	301	219	253	922	751	137	84
934	256	459	421	140	64	214	953	859
426	197	42	761	258	282	248	122	420
285	634	649	938	655	154	75	605	811
448	115	731	443	87	93	276	156	241
336	177	210	174	110	801	8	498	890
798	978	281	20	864	362	479	543	412
924	441	339	323	672	915	588	258	743
725	704	600	355	334	13	433	580	665
893	130	846	370	38	28	869	29	994
800	417	402	224	219	392	159	610	527
732	729	742	233	49	897	913	668	436
759	654	469	368	897	93	566	305	44
414	725	463	993	95	465	716	6	772
551	363	274	770	974	200	564	957	91
420	128	549	370	88	551	283	205	671
765	143	420	421	345	248	138	46	290
797	778	280	375	535	113	784	97	588
847	370	440	383	885	573	595	608	119
629	152	624	612	759	771	690	896	295
363	288	943	34	903	268	130	710	840
679	598	985	887	874	631	931	963	742
157	775	394	507	520	165	197	986	803
556	643	607	859	959	716	16	295	940
786	48	360	561	895	20	723	717	338
673	960	981	106	93	846	59	402	555
148	996	788	730	357	177	209	94	194
754	91	478	119	137	734	551	81	841
691	159	376	611	501	152	698	603	37
263	970	887	626	926	669	985	401	731
53	325	678	160	634	362	444	930	570
759	844	7	713	622	707	741	79	255
98	386	621	257	644	652	848	257	21
883	703	733	188	515	815	505	565	333
11	359	881	340	93	407	476	743	607
215	293	688	164	746	634	529	708	599
778	845	95	613	91	652	35	174	440
258	556	550	299	411	527	392	630	778
59	178	566	159	261	626	994	988	438

QAS-026

YOUR COMPANY STANDARD OPERATING PROCEDURE

SOP No. QAS-026.00 Effective date: mm/dd/yyyy
Approved by:

TITLE: **Physical Characteristics Check of Product and Foil during Sealing**

AUTHOR: ____

Name/Title/Department

____

Signature/Date

CHECKED BY: ____

Name/Title/Department

____

Signature/Date

APPROVED BY: ____

Name/Title/Department

____

Signature/Date

REVISIONS:

No.	Section	Pages	Initials/Date

SOP No. QAS-026.00 Effective date: mm/dd/yyyy
Approved by:

Subject: Physical Characteristics Check of Product and Foil during Sealing

45.1 PURPOSE

To ensure that the characteristic attributes of product and foil meet the acceptance criteria.

45.2 RESPONSIBILITY

It is the responsibility of the QAI to follow the procedure. The QA systems and inspection manager is responsible for the SOP compliance.

45.3 CHECKS DURING SEALING OPERATION

45.3.1 Attributes of Product and Foil Checks

1. Check the strips/blisters coming out from the machine on an hourly basis for physical attributes.
2. In blister machines, check tablets/capsules in the hopper/on the disk/on the conveyor for physical attributes. If any broken, chipped, or other defective tablets/capsules are found in the hopper/disk or conveyor, inform the operator and packaging manager and ask to sort out the defective tablets/capsules. Also, visually inspect the tablets in the machine channels for broken, chipped, or spotted tablets.
3. Collect 10 strips/blisters from the machine during sealing for attribute checks. Check for the following:
   i. Proper cutting
   ii. Proper overlapping of printed aluminum foil and plain PVC foil in strips/blisters
   iii. Wrinkle punctures
   iv. False sealing (open)
   v. Broken capsules/tablets in each pocket
   vi. Double tablets in each pocket
   vii. Clearly visible channels or cut on pocket borders
   viii. Misprint on the printed side of strips/blisters
   ix. Correct counts
   x. Check manufacturing date, expiry date, and batch number for correctness and legible overprinting
   xi. Empty pocket
4. For empty pockets, collect samples randomly from the lot stripped/blistered during last 2 hours as per the plan attached (Single Sampling, Tighten inspection Mil-Std 105E Level 1. Table II-B; Annexure II).
5. Check each strip/blister pocket visually and decide the disposition as per attached acceptance plan (Annexure I) based on Mil-Std 105E (Color/opaque PVC).
6. If the lot fails, ask for 100% inspection and collect the samples again randomly from the lot inspected and decide the disposition as per the acceptance plan.
7. After completion of the batch, randomly collect samples from the complete lot and check for the empty pockets and decide disposition as per the attached acceptance plan (Annexure I).
8. If the lot fails then inform the packaging manager for 100% inspection.
9. After 100% inspection, collect samples as per the attached sampling plan (Annexure I) and decide the disposition.

SOP No. QAS-026.00 Effective date: mm/dd/yyyy
Approved by:

45.3.2 Leak Test

1. Carry out the leak test at the start of the sealing operation on 10 blisters/strips coming out from sealing blister machines.
2. If the leak test fails, all the strips/blisters sealed by that time should be destripped/deblistered by the packaging department.
3. If the problem occurs continuously, then repeat the leak test till the sealing is satisfactory.
4. Carry out the leak test during and at the end of the strip/blister sealing process by collecting samples randomly from the lot stripped/blistered.
5. Check each strip pocket visually and decide the disposition as per the attached acceptance plan of reduced inspection.
6. If the lot fails in the leak test, ask for 100% inspection and collect the samples again randomly from the inspected lot and decide the disposition as per "A. Normal Inspection—Single Sampling Plan"; Refer to Annexure III.
7. Inform the packaging manager immediately in case of any anomaly or deviation.
8. Perform the leak test as per SOP No. QAS-000.

45.4 REASON FOR REVISION

First time issued for the ABC Pharmaceutical Company.

45.5 ATTACHMENT FORMS

ABC PHARMACEUTICAL COMPANY

SOP No.: QAS-026.00
Issued date:

Annexure I

A. Normal Inspection—Single Sampling Plan (Based on Mil-Std 105 E. Table 11-A)

			AQL 0.65	
Lot Size Unit Strip	Code (Special Level) (S-1)	Sample Size (*n*) (Unit Strip)	Ac (Unit Pocket)	Rej (Unit Pocket)
2–8	A	2	0	1
9–15	A	2	0	1
16–25	A	2	0	1
26–50	A	2	0	1
51–90	B	3	0	1
91–150	B	3	0	1
151–280	B	3	0	1
281–500	B	3	0	1
501–1200	C	5	0	1
1201–3200	C	5	0	1
3201–10,000	C	5	0	1
10,001–35,000	C	5	0	1
35,001–150,000	D	8	0	1

ABC PHARMACEUTICAL COMPANY

SOP No.: QAS-026.00
Issued date:

Annexure II

Sampling and Acceptance Plan for Empty Pockets of Strips/Blisters
Tightened Inspection—Single Sampling
(Based on Mil-Std 105E. Table 11-B)

Lot Size Unit Strip	Code (Level – I)	Sample Size (*n*)	AQL 0.065	
			Ac	Rej
501–1200	G	32	0	1
1201–3200	H	50	0	1
3201–10,000	J	80	0	1
10,001–35,000	K	125	0	1
35,001–150,000	L	200	0	1

ABC PHARMACEUTICAL COMPANY

SOP No.: QAS-026.00
Issued date:

Annexure III

B. Reduced Inspection—Single Sampling Plan (Based on Mil-Std 105E. Table 11-C)

Lot Size Unit Strip	Code (Special Level) (S-1)	Sample Size (*n*) (Unit Strip)	AQL 0.65	
			Ac (Unit Pocket)	Rej (Unit Pocket)
2–8	A	2	0	1
9–15	A	2	0	1
16–25	A	2	0	1
26–50	A	2	0	1
51–90	B	2	0	1
91–150	B	2	0	1
151–280	B	2	0	1
281–500	B	2	0	1
501–1200	C	2	0	1
1201–3200	C	2	0	1
3201–10,000	C	2	0	1
10,001–35,000	C	2	0	1
35,001–150,000	D	3	0	1

QAS-027

YOUR COMPANY STANDARD OPERATING PROCEDURE

SOP No. QAS-027.00 Effective date: mm/dd/yyyy
Approved by:

TITLE: **Raw Material QA Audit Checks**

AUTHOR: ______________________________

Name/Title/Department

____

Signature/Date

CHECKED BY: ______________________________

Name/Title/Department

____

Signature/Date

APPROVED BY: ______________________________

Name/Title/Department

____

Signature/Date

REVISIONS:

No.	Section	Pages	Initials/Date

SOP No. QAS-027.00 Effective date: mm/dd/yyyy
Approved by:

Subject: Raw Material QA Audit Checks

46.1 PURPOSE

The purpose of this SOP is to provide a procedure for audit checks of dispensing raw material.

46.2 RESPONSIBILITY

The QAI will be responsible for following the procedure and the QA manager will be responsible for the SOP compliance. The QA director is responsible for the overall system compliance.

46.3 PROCEDURE

Computer documentation system is implemented for ensuring

a. FIFO (first-in first-out system of raw materials)
b. Check for issuance of correct quantities
c. Location of a particular lot
d. Inventory status of a particular lot of material

46.3.1 Raw Material QA Audit Checks

1. The QAI has to confirm that
   a. The issuance is according to FIFO.
   b. All the entries on the computer are present and correct.
2. The QAI has to confirm that the quantities dispensed are correct and as per the master manufacturing formula requisition.
3. The balance calculation is correct and tallies with the physical stock of the lot of the corresponding raw material.
4. In case of any discrepancy, investigate the cause of shortage or excess.
5. The weighing area manager will raise an inventory exception report for the accounts department and the production director in case of physical shortage or excess of a raw material.
6. In cases where the in-use lot is quarantined (QA/QC HOLD), the next lot could be used and if the quarantined lot is approved by QA/QC at a later time, then discontinue the use of the lot taken after that and start using the older lot again with the label of *Re-Released ON*; date is mentioned in the system.
7. The raw material of the next lot (which was in use) will have a notation "hold as previous lot is now released."
8. In case the lot is rejected, the computer dispensing/inventory system should have the remarks or notation "rejected on _______."
9. The QAI will check the balance quantity at the time of dispensing.

46.3.2 GMP Checks

The area QAI (weighing area) will check the general GMP of the weighing area daily as per SOP QAS-000 and upon finding any anomaly, will raise the deviation citation format and will follow the corrective action procedure.

SOP No. QAS-027.00 Effective date: mm/dd/yyyy
Approved by:

46.3.3 Area Clearance Check

At the start of each weighing of a batch, the area QAI (weighing area) will make sure that all the requirements of area clearance are being followed as per the area clearance format and signed by the weighing area supervisor/officer and also the weighing area manager.

If all the requirements are met, then the area QAI will countersign the weighing area equipment clearance (Annexure I).

46.4 REASON FOR REVISION

First time issued for the ABC Pharmaceutical Company.

46.5 ATTACHMENT FORM

ABC PHARMACEUTICAL COMPANY

LEVEL 2
SOP No.: QAS-
Issued on:
Revision No.:
Initiator name:

ANNEXURE I
DRUG WEIGHING AREA
EQUIPMENT CLEARANCE CHECKLIST

Product Name: ______________________
Code No.: ______________________
Batch No.: ______________________

Weighing to be carried out in the following booth(s):

1. [×]
2. [×]
3. [✓]

The areas have been cleared in accordance with the checklist	☐
Weighing area is clean	☐
Balance/scales in equilibrium	☐
Balances are calibrated	☐
No raw material of previous batch in weighing booth	☐
Scoops dry and clean	☐
Only one raw material at a time is brought for weighing	☐
Operator wearing caps, gloves, and mask	☐

Signature and date	Signature and date	Signature and date
______________	______________	______________
Weighing area supervisor	**QA in-process**	**Weighing area manager**

QAS-028

YOUR COMPANY STANDARD OPERATING PROCEDURE

SOP No. QAS-028.00 Effective date: mm/dd/yyyy
Approved by:

TITLE: **Incidence Citation Report**

AUTHOR: ____________________________________

Name/Title/Department

____

Signature/Date

CHECKED BY: ____________________________________

Name/Title/Department

____

Signature/Date

APPROVED BY: ____________________________________

Name/Title/Department

____

Signature/Date

REVISIONS:

No.	Section	Pages	Initials/Date

SOP No. QAS-028.00 Effective date: mm/dd/yyyy
Approved by:

Subject: Incidence Citation Report

47.1 PURPOSE

The purpose of this SOP is to describe the procedure and events to raise an incident report and to initiate the corrective actions in the event of cGMP violations and unusual events.

47.2 RESPONSIBILITY

It is the responsibility of all employees to follow the procedure of generating incidence citation report (ICR) and take corrective actions. The area supervisor and the manager will be responsible for completing the corrective action in case of an incidence reporting (where required). The QA and manufacturing directors will be responsible for the SOP compliance.

47.3 PROCEDURE

1. ICR (Attachment I) will be raised in cases where an unusual event or cGMP violation and observation has been made by an employee, analyst, inspector, or others.
2. ICR will be forwarded to the QA manager for logging the incidence report.
3. Separate registers will be maintained for logging, registering, and assigning the ICR in a sequential number.
4. The ICR will be marked back to the concerned department for following up and reporting of corrective actions completion.
5. The completed ICR will be returned to the QA manager to be signed off by the QA director for close-out and filing for reference in a separate folder.

47.3.1 cGMP Violation

1. In the event of cGMP violations, the QAI will raise the ICR and corrective action will be taken on spot before further continuation of work.
2. The concerned manager will be responsible for the implementation of the suggested corrective action and execution of the preventive measure/s which may include but are not limited to retraining of the identified personnel, disciplinary action including issuance of warning letter, verbal reprimand, transfer, or even termination if required.

   The corrective and preventive measure will be prescribed individually depending upon the causes of the incidence and severity of the citation, explanation of the personnel and supervisor level involved, recurrence, and so on.
3. Besides, QA will also conduct monthly trending of the ICR of the plant to keep the management informed and initiate necessary actions (refer to Attachment II).

47.3.1.1 Classification of Incidence Citation

Critical: An incidence that, by judgment, experience, and situation indicate if happens, would cause an unsafe condition in the product or may prevent performance/functionality of the product.

Major: An incidence that may result in defect/s in the product reducing the usability of the product.

Minor: An incidence that, indicated by experience, situation, and judgment, may not result in a product with reduced usability but is a departure from established practices or standards.

SOP No. QAS-028.00 Effective date: mm/dd/yyyy
Approved by:

47.3.1.2 Unusual Incidence or Event Reporting

1. All those citation which cannot be categorized as a violation of cGMP (refer to GMP SOP QAS-023) are unintentional and fall under the group of "nonconformance to normal practices" or "out of the usual run of things" will also be reported on the ICR.
2. The ICR will be raised by manufacturing, packaging, quality assurance, and quality control in case of an event unusual to normal practice.
3. Some examples include
   a. Breakdown of machine during processing
   b. Batch/lot complying after double sampling or resampling, and so on
   c. Pest control
      i. If the pest is observed in the area by any employee, the information should be given to the supervisor.
      ii. The supervisor of the area should stop the operation (manufacturing and/or packaging). The QAIs are also authorized to stop the line if any pest is observed in the area (separate SOP for corrective/preventive measures in case of pest observance in aseptic core).
      iii. The product shall be protected as per the general production system.
      iv. The area shall be locally treated to spot the insect and remove it.
      v. The area must be physically inspected for the absence of pest by the area supervisor and the QAI.
      vi. The area will be given line clearance.
      vii. The QAI will raise the ICR (refer to Attachment I) and will document the corrective action.
      viii. The product manufactured or packed before the incident must be inspected 100% before process continuation by the production/packaging supervisor as applicable.

47.4 DOCUMENTATION

- ICR (Attachment I)
- Monthly trend report format (Attachment II)

47.5 REASON FOR REVISION

First time issued for the ABC Pharmaceutical Company.

47.6 ATTACHMENT FORMS

ABC PHARMACEUTICAL COMPANY

LEVEL 3
Attachments No.: QAS-
Issued on:
Revision No.:

Attachment I
Incidence Citation Report and Corrective Action Request

ICR No.: ______________

<table>
<tr><td rowspan="3">INITIATOR</td><td>CITATION
Written by: ______________ Designation: __________ Initial: ______________
Through (department head): ______________________ Date: ______________</td></tr>
<tr><td>Nature of citation: ☐ CRITICAL ☐ MAJOR ☐ MINOR
Description of event: ______________________________</td></tr>
<tr><td>Recommendations: ______________________________</td></tr>
<tr><td>QAD</td><td>Corrective action request (CAR) to be completed by: Date: ______________
Signature QAD: ______________</td></tr>
<tr><td>DEPARTMENT HEAD INVOLVED</td><td>Corrective action implementation report: CAR completed: ☐ YES ☐ NO
Remarks or explanation: ______________________________
Signature of department head: ______________________ QAIs: ______________</td></tr>
</table>

ABC PHARMACEUTICAL COMPANY

LEVEL 3
Attachments No.: QAS-
Issued on:
Revision No.:

Attachment I
Incidence Citation Report and Corrective Action Request

CLOSE-OUT	CAR completion and close-out: ______________________________ **Quality assurance director**

cc: Initiator/Departmental head/QA manager

LEVEL 3
Attachments No.: QAS-028-II
Issued on:
Revision No.: 6

Attachment II
TREND ANALYSIS
INSPECTION REPORT/CORRECTIVE ACTION

To:
From: QA manager
Date: **Plant:** **Area:**

S. No	OBSERVATION/ DEVIATION	TYPE OF DEVIATION	NUMBER OF DEVIATON FOR THE MONTH	TOTAL DEVIATION YTD	TOTAL OBSERVATION YTD	JANUARY	FEBUARY	MARCH	APRIL	MAY	JUNE	JULY	AUGUST	SEPTEMBER	OCTOBER	NOVEMBER	DECEMBER	DEVIATION BY
01	**GENERAL GMP VIOLATION**																	
	Unattended operations																	
	Not wearing mask																	
	Door kept open during processing																	
	Dust/dirt in the area																	
	Improper clothing																	
	Dirty gowning																	
	Broken apparatus																	
	Unclean equipment																	
	Cages dirty/pallets dirty																	
	Material not on the pallet																	
	Cages unlocked																	
	Insects in production area																	
	Long hair																	
	Long nails																	
	Wearing jewelry																	

Page 1 of 5

LEVEL 3
Attachments No.: QAS-028-II
Issued on:
Revision No.: 6

Attachment II

TREND ANALYSIS

INSPECTION REPORT/CORRECTIVE ACTION

To:
From: QA manager
Date: **Plant:** **Area:**

S. No	OBSERVATION/ DEVIATION	TYPE OF DEVIATION	NUMBER OF DEVIATON FOR THE MONTH	TOTAL DEVIATION YTD	TOTAL OBSERVATION YTD	JANUARY	FEBUARY	MARCH	APRIL	MAY	JUNE	JULY	AUGUST	SEPTEMBER	OCTOBER	NOVEMBER	DECEMBER	DEVIATION BY
	Mishandling/wastage of material																	
	Two processes at a time																	
	Unapproved/rejected mat usage																	
	Area cracked (floor, wall, & roof)																	
	Units out of limit (less or more)																	
	Rechecking due																	
02	**LINE CLEARANCE VIOLATION**																	
	Improper documentation																	
	Improper labeling																	
	No line clearnace																	

LEVEL 3
Attachments No.: QAS-028-II
Issued on:
Revision No.: 6

Attachment II

TREND ANALYSIS

INSPECTION REPORT/CORRECTIVE ACTION

To:
From: QA manager
Date: **Plant:** **Area:**

S. No	OBSERVATION/ DEVIATION	TYPE OF DEVIATION	NUMBER OF DEVIATON FOR THE MONTH	TOTAL DEVIATION YTD	TOTAL OBSERVATION YTD	JANUARY	FEBUARY	MARCH	APRIL	MAY	JUNE	JULY	AUGUST	SEPTEMBER	OCTOBER	NOVEMBER	DECEMBER	DEVIATION BY
	Other product/ commodities																	
	Documents in the area																	
	Antibiotic in other area or viceversa																	
03	**ENVIRONMENTAL/MICROBIOLOGICAL**																	
	RH out of limit																	
	Tempreture out of limit																	
	No fumigation																	
	Pressure out of limit																	
04	**RELATED TO DOCUMENT**																	
	Wrong entries																	
	No entries																	
	Incomplete documentation																	

LEVEL 3
Attachments No.: QAS-028-II
Issued on:
Revision No.: 6

Attachment II

TREND ANALYSIS

INSPECTION REPORT/CORRECTIVE ACTION

To:
From: QA manager
Date: **Plant:** **Area:**

S. No	OBSERVATION/ DEVIATION	TYPE OF DEVIATION	NUMBER OF DEVIATON FOR THE MONTH	TOTAL DEVIATION YTD	TOTAL OBSERVATION YTD	JANUARY	FEBUARY	MARCH	APRIL	MAY	JUNE	JULY	AUGUST	SEPTEMBER	OCTOBER	NOVEMBER	DECEMBER	DEVIATION BY
05	**RELATED TO LABELING**																	
	Improper labeling (area/ product)																	
	Not clear B.No/Exp.Date/ Mfg.Date																	
	Without label or status																	
	No QA release label																	
06	**REGARDING CALIBRATIONS**																	
	Daily calibration not updated																	
	Calibration due not performed																	
07	**LOG BOOK ENTERIES / CHECKING**																	
	Incomplete																	
	a. Equipment usage record																	
	b. Fumigation record																	

LEVEL 3
Attachments No.: QAS-028-II
Issued on:
Revision No.: 6

Attachment II
TREND ANALYSIS
INSPECTION REPORT/CORRECTIVE ACTION

To:
From: QA manager
Date: **Plant:** **Area:**

S. No	OBSERVATION/ DEVIATION	TYPE OF DEVIATION	NUMBER OF DEVIATON FOR THE MONTH	TOTAL DEVIATION YTD	TOTAL OBSERVATION YTD	JANUARY	FEBUARY	MARCH	APRIL	MAY	JUNE	JULY	AUGUST	SEPTEMBER	OCTOBERE	NOVEMBER	DECEMBER	DEVIATION BY
08	INAPPROPRIATE STORAGE																	
	Raw materials																	
	In-process material																	
	Bulk products																	
09	NONCOMPLIANCE TO / OF																	
	MFM																	
	SOP																	

SIGNATURE: ____________________

Quality assurance director

Page 5 of 5

QAS-029

YOUR COMPANY STANDARD OPERATING PROCEDURE

SOP No. QAS-029.00 Effective date: mm/dd/yyyy
Approved by:

TITLE: **Technical Documents Distribution Control**

AUTHOR: ____________________________________

Name/Title/Department

____

Signature/Date

CHECKED BY: ____________________________________

Name/Title/Department

____

Signature/Date

APPROVED BY: ____________________________________

Name/Title/Department

____

Signature/Date

REVISIONS:

No.	Section	Pages	Initials/Date

SOP No. QAS-029.00 Effective date: mm/dd/yyyy
Approved by:

Subject: Technical Documents Distribution Control

48.1 PURPOSE

The purpose of this SOP is to describe the procedure for the distribution of technical documents in the ABC Pharmaceutical Company.

48.2 RESPONSIBILITY

It is the responsibility of all departmental managers to follow the procedure. The assistant manager systems is responsible for the SOP compliance.

48.3 PROCEDURE

1. All original technical documents of the company will be kept under the control of the systems department and the PDL.
2. The technical documents are considered level 2 documents and cannot be copied by anyone except the systems department designate.
3. The technical documents include

Documents	Responsible
i. Manufacturing formula and method	R&D lab
ii. Finished production specification (FPS)	Systems
iii. Standard test method (STM)	Systems
iv. Stability report	Systems
v. Master packaging instruction (MPI)	R&D lab
vi. Packaging material specifications	R&D lab
vii. Raw material specification (RM)	Systems
viii. Test method validation report	Systems
ix. Registration documents	Systems
x. In-process control specification (mfg. and pack.)	Systems
xi. Raw material purchasing specification	R&D lab
xii. Packaging material purchase specification	R&D lab

4. All technical documents will have a fixed format describing information such as
   i. Written by
   ii. Reviewed by or verified by
   iii. Checked by
   iv. Approved by
5. After approval from the QA director, the master documents will be handed over to the systems department designated person by the concerned department through the "document update request" form.
6. The systems department designate will make the required number of photocopies as per the distribution list from the original document and stamp it in red ink as *not to be reproduced or copied.*
7. The manager systems/designate will be responsible for providing the current in use copy to the related department as per the distribution list.

SOP No. QAS-029.00 Effective date: mm/dd/yyyy
Approved by:

8. The systems department will maintain the log book for the distribution of each document type, which will be signed and dated by the recipient of the document.
9. Superseded copies of the technical documents will be returned to the systems department for destruction.
10. Each department will be responsible for maintaining the current and updated document files concerning them and marked to them.
11. Master technical documents files will be maintained by the systems department.
12. Master technical void documents will be kept for indefinite time and to be stamped *void* in red ink.
13. A copy of the approved technical document will be forwarded with attachment number 1 to the concerned manager by the manager systems to conduct the training of the personnel involved in executions of the document. After training, the record will be returned to the systems department and will be kept with the master document.

48.4 REASON FOR REVISION

First time issued for the ABC Pharmaceutical Company.

48.5 ATTACHMENT FORMS

ABC PHARMACEUTICAL COMPANY

Annexure-I
SOP No.: QAS-
Issued on:
Revision No.:
Initiator name:

TECHNICAL DOCUMENTS TRAINING RECORD

i. Manufacturing formula and method	☐	vii. Raw materials specifications	☐
ii. Finished production specifications (FPS)	☐	viii. Test method validation report	☐
iii. Standard test method (STM)	☐	ix. Registration documents	☐
iv. Stability report	☐	x. In-process control specification (mfg. and pack)	☐
v. Master packaging instruction (MPI)	☐	xi. Raw material purchasing specification	☐
vi. Packaging material specifications	☐	xii. Packaging material purchase specification	☐

Trainees (Name)	ID No.	Signature of Understanding	Date	Evaluation	Results	Training Conducted by	Signature of the Trainer

Evaluation: O = oral
R = written
Q = questionnaire
E = explained on site

Results: M = marginal (re-train required)
A = average
S = satisfactory
C = commendable

Attachment No.: 1
SOP No.: QAS-
Issued on:
Revision No.:
Initiator name:

ADDITIONAL COPIES (ELECTRONIC/HARD) REQUEST OF TECHNICAL DOCUMENTS

To: Section Head Systems

From: ______________________________

Date: ______________________________

Document ID No.: ____________________

Document code No.: __________________

Version/revision No.: ________________

Issued on: ___________________________

No. of copies required: _______________

Copies required for use: | **In the company** | **Out of company** |

Registration purpose in ________ ________________________ **(Country)**

Requested from ____

Last version of document sent ________________________________

Please enclose received request, E-mail, memo, etc.

_________________________ _________________________

Departmental head approval **Quality assurance director approval**

SYSTEM DEPARTMENT USE

Request received on: ____________ **By** ___________________

Copy provided on: ______________ **By** ___________________

Mark of stamp used: ____

QAS-030

YOUR COMPANY STANDARD OPERATING PROCEDURE

SOP No. QAS-030.00 Effective date: mm/dd/yyyy
Approved by:

TITLE: **Key Performance Indicator**

AUTHOR: ______________________________

Name/Title/Department

____

Signature/Date

CHECKED BY: ______________________________

Name/Title/Department

____

Signature/Date

APPROVED BY: ______________________________

Name/Title/Department

____

Signature/Date

REVISIONS:

No.	Section	Pages	Initials/Date

SOP No. QAS-030.00 Effective date: mm/dd/yyyy
Approved by:

49.1 PURPOSE

To provide a written procedure for documenting key performances, milestones, achievements, incidents, accomplishments, problems, and major deviations that took place in various departments of the ABC Pharmaceutical Company.

49.2 RESPONSIBILITIES

1. The QA officer will be responsible to initiate the report and send a request to the managers/directors of all the departments to provide the key performance indicators (KPIs) of the current month.
2. Managers/directors of the corresponding departments will be responsible to send the KPIs of their respective departments to the QA officer by the 5th day of each month.
3. All departmental directors are responsible for system compliance.

49.3 PROCEDURE

Each departmental manager/director will fill out the template given in the following pages according to the requirements of their respective department and will forward it to the QA officer no later than the 5th day of each month.

49.4 ATTACHMENT FORMS

49.4.1 Month–Year

49.4.1.1 QA Manager (Solid/Liquid/Sterile/Powder)

Key performance indicator: INTERNAL AUDITS FOLLOW UP (PENDING ISSUES)

Report for the month of ....

Area	Year	Observations	Corrective Actions Completed	Pending CARs	Remarks
Production (solid)					
Packaging (solid)					
Production (liquid)					
Packaging (liquid)					
Production (sterile)					
Packaging (sterile)					
Production (powder)					
Packaging (powder)					
Stores					
Maintenance					
Quality control					
Research & development					

Information provided by: ______________________

Name and signature: ______________________

Date: ______________________

In the above template, the QA manager of the corresponding area will fill out the information related to the internal audit conducted by the QA department.

Key performance indicator: EXTERNAL & ISO AUDITS FOLLOW UP (PENDING ISSUES)

Report for the month of ....

Audit Conducted By[1]	Date	Observations[2]	Corrective Actions Completed[3]	Pending CARs[4]	Department[5]	Remarks[6]

Information provided by: ___________________________

Name and signature: ______________________________

Date: ____

This template will provide information associated with the audits conducted by any external agency, buyers, regulatory bodies, or International Standards for Organization.

Note:

1. Enter here the name of the regulatory body or customer, conducting the audit.
2. If the auditors have given observations/deviations observed.
3. If the corrective actions have been completed.
4. Indicate here the remaining corrective actions report.
5. Name of the department for which the observation was made.
6. Any remarks by the department or the auditors.

Key performance indicator: VALIDATION ACTIVITIES

Report for the month of ....

S/No.	Description[1]	Protocol		Executed Total No.[4]	Not Executed[5]	Remarks/Corrective Measure[6]
		Written[2]	Executed[3]			

Information provided by: ___________________________

Name and signature: ___________________________

Date: ___________________________

Note:

1. This column will provide description of the validation, for example, a new product manufacturing process, new equipment, facility and change in the supplier of API, and so on.
2. Check the column if the protocol has been written and approved.
3. Check the column if the protocol has been executed.
4. Number of protocols executed during the month.
5. Number of protocols that could not be executed.
6. Add here if any comment regarding the validation activity is required.

Key performance indicator: MARKET COMPLAINTS (VALID/INVALID) AND RECALLS

Report for the month of ....

Kindly provide the required information as per the following data sheet up to the 5th day of each month.

Additional sheets may be used for details wherever necessary.

S/No.	Product Name[1]	Batch No.[2]	Nature of Complaint[3]	Complaint from[4]	Response/ Corrective Measure[5]	Valid/ Invalid[6]

Information provided by: ______________________

Name and signature: __________________________

Date: ______________________________________

Note:

1. Name of the product for which complaint is received.
2. Batch number of the product.
3. Nature of complaint, for example, the description of complaint will be entered in this column.
4. In this column, the name of complainant will be entered, for example, if it is from a customer, from health regulatory laboratory, or from a pharmacy store.
5. Details of the corrective measures taken.
6. Conclusion, if the complaint was genuine or not.

Key performance indicator: IN-PROCESS NONCOMPLIANCE/PROCESS DEVIATION

Report for the month of ....

Product	Batch #	In-process Nonconformance[1]	Root Cause[2]	Corrective Action[3]	Preventive Action[4]	
						Please fill all columns and provide detailed information

Information provided by: ____________________________

Name and signature: ____________________________

Date: ____________________________

Note:

1. Nature of nonconformance as to which process does it belong to.
2. If the root cause has been determined, indicate in this column.
3. Indicate here what corrective action has been taken.
4. Write in this column if and what preventive action will be taken to avoid recurrence.

Key performance indicator: REPROCESSING/REWORK/SORTINGS (PRODUCTION)

Report for the month of ....

Kindly provide the required information as per the following data sheet up to the 5th day of each month.

Additional sheets may be used for details wherever necessary.

Product[1]	Batch No.[2]	Rework/ Sorting[3]	Root Cause[4]	Corrective Action[5]	Preventive Measure[6]	Please fill all columns and provide detailed information

Information provided by: ______________________

Name and signature: ______________________

Date: ______________________

Note:
1. Name of the product.
2. Batch number of the product.
3. Whether it was reworking or sorting incident.
4. Root cause determined for the incident.
5. Enter here if a corrective action has been taken.
6. What preventive measures were taken?

Key performance indicator: GMP, GLP, & SOP DEVIATION (ICRs)

Report for the month of ....

Total number of ICRs: ....

Deviation from GMP & SOP[1]	Minor[2]	Major[3]	Critical[4]	Total No.[5]	Root Cause[6]	Corrective Action[7]	Preventive Action[8]
Stores							
Production							
Quality control							
Packaging							
Maintenance							

Information provided by: ___________________________

Name and signature: ________________________________

Date: ____

Note:

1. The corresponding areas have already been given in this column for which the deviation is to be recorded.
2. If the deviation is minor.
3. If the deviation is major.
4. If the incident is critical.
5. Total number of deviations in the corresponding area.
6. Root cause determined.
7. What corrective action was taken, if any.
8. What preventive measures were recommended, if any.

Key performance indicator: CHANGE CONTROL SUMMARIES FOR THE MONTH

Report for the month of ....

Information provided by: ______________________

Name and signature: __________________________

Date: ______________________________________

Key performance indicator: AIR BORNE PARTICULATE MONTHLY SUMMARY (STERILE PLANTS)

Report for the month of ....

Cleanness Class[1]	No. of Test Performed[2]	Out of Specification[3]	Total[4]	Corrective Measure[5]

Information provided by: ____________________

Name and signature: ____________________

Date: ____________________

Note:

1. The class of the room, for example, class 100, 1000, or 100,000.
2. Test performed for the air-borne particles.
3. If the result is out of specification.
4. Total number of particles.
5. Indicate in this column what corrective measure was taken.

Key performance indicator: ALL DEPARTMENTS SIGNIFICANT EVENTS

Report for the month of..........................

Kindly provide the required information as per the following data sheet. Additional sheets may be used for details wherever necessary.

1
2
3
4
5
6
7
8
9
10
11
12
13
14
15
16

Information provided by: ____________________

Name and signature: ____________________

Date: ____________________

Key performance indicator: INCOMING MATERIALS (RAW MATERIALS) NUMBER OF LOTS WITH / WITHOUT LOT NUMBERS

Report for the month of ....

Total lots received: ____________________ Nos.

With lot number: ____________________ Nos.

Without lot number: ____________________ Nos.

Details of Raw Materials without Lot Number

S/No.	Material Name	Supplier

Information provided by: ____________________

Name and signature: ____________________

Date: ____________________

Key performance indicator: SOPs WEBSITE STATUS, SOPs REVISED/NEW CANCELLED

Report for the month of ....

SOP Website Status[1]	No. of SOPs Revised[2]	No. of New SOPs[3]	No. of SOPs Cancelled[4]	Remarks[5]

Information provided by: ________________________

Name and signature: ___________________________

Date: ____

Note:

1. State here the status of the website of the SOPs on intranet, if the website is under maintenance or reconstruction etc.
2. Total number of SOPs revised in the company.
3. How many new SOPs included in the system.
4. How many SOPs were discarded.
5. Remarks, if any.

Key performance indicator: TEMPERATURE MONITORING FOR SHIPPING (FINISHED PRODUCT)

Report for the month of ....

Shipping Route[1]	No. of Shipments[2]	Recommended Temperature[3]	Temperature within Limit[4]	Deviation from Recommended Temperature[5]	Remarks[6]

Information provided by: ______________________

Name and signature: ______________________

Date: ______________________

Note:

1. Country to which the material is shipping to.
2. Total number of shipments to this consignee.
3. Temperature of the trailer recommended for each type of products in the consignment.
4. Check if the temperature remained within limits during the transportation.
5. Report in this column if any deviation occurred in temperature during transportation.
6. Remarks, if any.

Key performance indicator: INCOMING MATERIALS (RAW & PACKAGING)

Report for the month of ....

Total raw material received: ______________________________ Nos.

Total packaging materials received: ______________________________ Nos.

Total raw material resampled: ______________________________ Nos.

Total raw material retested: ______________________________ Nos.

Total raw material received with import Permit: ______________________________ Nos.

Details of Raw Materials Received with Import Permit (Ministry of Health Approval)

S/No.	Material Name	Supplier

Information provided by: ________________________

Name and signature: ____________________________

Date: ____

49.4.1.2 Production Manager (Solid/Liquid/Sterile/Powder/Biotechnology Plant)

Key performance indicator: TOTAL NUMBER OF BATCH MANUFACTURED FOR EACH DOSAGE FORM

Report for the month of ....

Dosage Forms	Total No. of Batches	Total (units/numbers)
Tablets		
Caplets		
Capsule		
Granules		
Syrup		
Drops		
Suspension		
Cream and ointment		
Injectable		
Powder filling		

Information provided by: ______________________

Name and signature: ______________________

Date: ______________________

Key performance indicator: IN-PROCESS NONCOMPLIANCE/PROCESS DEVIATIONS

Report for the month of ....

Product	Batch #/Number of Batches[1]	In-process Nonconformance[2]	Root Cause[3]	Corrective Action[4]	Preventive Action[5]	Please fill all columns and provide detailed information

Information provided by: ______________________

Name and signature: ______________________

Date: ______________________

Note:

1. Total number of batches manufactured in the respective dosage form area.
2. Number of nonconformances occurred during the processes.
3. If the root cause of the nonconformance was determined.
4. Brief description of corrective actions taken.
5. Brief description of preventive actions taken.

Key performance indicator: REPROCESSING/REWORK/SORTINGS (PRODUCTION SOLID DOSAGE)

Report for the month of ....

Product[1]	Batch No.[2]	Rework/ Sorting[3]	Root Cause[4]	Corrective Action[5]	Preventive Measure[6]	Please fill all columns and provide detailed information

Information provided by: ________________________

Name and signature: ____________________________

Date: ____

Note:

1. Name of product.
2. Batch number of the product.
3. Nature of rework involved.
4. Reason/root cause for re-working.
5. Brief description of the corrective action taken.
6. Brief description of the preventive action taken.

Key performance indicator: DEVIATION FROM STANDARD YIELD (BULK)

Report for the month of ....

Bulk Product[1]	Batch No.[2]	Date	Standard Yield[3]	Actual[4]	Reason[5]

Information provided by: ________________________

Name and signature: ____________________________

Date: ____

Note:
1. Name of the bulk product.
2. Batch number of the product.
3. Mention the actual yield of the batch as per master formula.
4. Actual yield of the batch received.
5. Brief description of probable reason as per investigation report.

Key performance indicator: ABNORMAL VARIANCE (FINISHED PRODUCT & PACKAGING)

Report for the month of ....

Finished Product[1]	Batch No.[2]	Date[3]	Standard Yield[4]	Actual[5]	Root Cause[6]	Corrective Action[7]	Preventive Action[8]

Information provided by: ____________________

Name and signature: ____________________

Date: ____________________

Note:

1. Name of the finished product.
2. Batch number of the product.
3. Date of the occurrence of the incident.
4. Enter here the standard yield as per master manufacturing formula or master packaging instructions.
5. Enter in this column the actual yield of product.
6. If the root cause has been determined, enter in this column.
7. What corrective actions were recommended?
8. What preventive actions were taken?

Key performance indicator: PROBLEMS DURING PACKAGING (COMMODITIES RELATED AND OTHER BREAKDOWNS)

Report for the month of ....

Product	Batch No.	Code No.	Commodity[1]	Nature of Problem[2]	Preventive Measure[3]	Closed Out[4]	
						Yes	No

Information provided by: ______________________

Name and signature: ___________________________

Date: ______________________________________

Note:

1. Type of packaging commodity associated with the problem or breakdown.
2. Exact nature of problem/incident.
3. If a preventive measure was taken.
4. Indicate in this column by checking the appropriate word YES or NO if the incident report was closed or still open.

Key performance indicator: TOTAL NUMBER OF BATCHES PACKED (STERILE)

Report for the month of ....

Dosage Forms	Total No. of Batches	Total (units/numbers)
Ampoules		
Vials (Liquid fill)		
Vials (Lyophilized)		

Information provided by: ________________________

Name and signature: ___________________________

Date: ____

Key performance indicator: TOTAL NUMBER OF BATCHES MANUFACTURED AND PACKED FOR EACH DOSAGE FORM (CEPHS. & PENS.)

Report for the month of..........................

Dosage Forms	Total No. of Batches	Total (units/numbers)
Tablets (Cephalosporin)		
Capsules (Cephalosporin)		
Capsules (Penicillin)		
Granules (Cephalosporin)		
Granules (Penicillin)		

Information provided by: ______________________

Name and signature: __________________________

Date: ______________________________________

Key performance indicator: TOTAL NUMBER OF BATCHES MANUFACTURED AND PACKED FOR EACH DOSAGE FORM (LIQUID)

Report for the month of ....

Dosage Forms	Total No. of Batches	Total (units/numbers)
Syrup		
Suspension		
Drops		
Cream		
Sterile cream		
Ointment		
Suppositories		
Ovules		
Spray/inhaler		
Sachets (liquid fill)		
Sachets (powder fill)		
Gel		
Solution		

Information provided by: ___________________________

Name and signature: ______________________________

Date: ____

Key performance indicator: TOTAL NUMBER OF BATCHES MANUFACTURED

Report for the month of ....

Dosage Forms	Total No. of Batches	Total (units/numbers)
Insulin		
Erythropoietin		

Information provided by: ______________________

Name and signature: ___________________________

Date: ______________________________________

49.4.1.3 Stores

Key performance indicator: INCOMING MATERIALS (RAW MATERIALS) NUMBER OF LOTS WITH/WITHOUT LOT NUMBERS

Report for the month of ____________________, **year** __________________

Total lots received: ____________________________________ Nos.

With lot number: ____________________________________ Nos.

Without lot number: ____________________________________ Nos.

Details of Raw Materials without Lot Number

S/No.	Material Name	Supplier

Information provided by: ______________________

Name and signature: __________________________

Date: _______________________________________

Key performance indicator: MATERIALS DESTRUCTION RECORD (RAW MATERIAL)

Report for the month of __________________, year_____________________

Total: ______________________________________ Nos.

Materials Destroyed

Serial No.	Raw Material Name	Lot No.	Quantity[1]	Reason[2]	Date of Destruction[3]

Information provided by: ____________________________

Name and signature: _________________________________

Date: ____

Note:

1. In this column total quantity of the corresponding material which is destroyed will be recorded.
2. Record the reason for destruction, for example, expired, failed to comply with specifications or not required to manufacture the finished product, and so on.
3. The date when the material is destroyed.

Key performance indicator: MATERIALS DESTRUCTION RECORD (FINISHED PRODUCT)

Report for the month of __________________, year____________________

Total: ______________________________________ Nos.

Materials Destroyed

Serial No.	Product Name	Batch No.	Quantity[1]	Reason[2]	Date of Destruction[3]

Information provided by: ____________________________

Name and signature: _________________________________

Date: ____

Note:

1. In this column total quantity of the corresponding material which is destroyed will be recorded.
2. Record the reason for destruction, for example, expired or failed to comply with specifications, and so on.
3. The date when the material is destroyed.

Key performance indicator: STATUS OF EXPIRED MATERIALS (FINISHED PRODUCTS)

Report for the month of _______________, **year**_________________

Materials	Total No. of Expired Lots	Quantity	Details (Batch/Lot No.) Reason, etc.

Information provided by: _______________________

Name and signature: ___________________________

Date: _______________________________________

Key performance indicator: STATUS OF EXPIRED RAW MATERIALS (RETESTING MATERIALS)

Report for the month of ________________, year__________________

Materials	Total No. of Expired Lots	Quantity	No. of Lots Retesting	Details (Batch/Lot No.) Reason, etc.

Information provided by: ________________________

Name and signature: ____________________________

Date: ____

49.4.1.4 Maintenance Managers (Production Plants, Packaging, Utilities, Electrical & Electronics, HVAC)

Key performance indicator: PREVENTIVE MAINTENANCE OF MACHINES & EQUIPMENTS AS PER SCHEDULE AND BREAK DOWN MAINTENANCE

Report for the month of _______________, year_________________

Planned Maintenance

Area	Number of Machines/ Equipment	Number of Maintenance Due or Planned	Maintenance Performed	Remarks/ Corrective or Preventive Measures

Breakdown Maintenance

Area	Number of Breakdowns	Nature of Breakdown	Corrective Action	Preventive Measures

Information provided by: _______________________

Name and signature: ___________________________

Date: ____

Key performance indicator: EQUIPMENT CALIBRATION

Report for the month of _______________, **year**_________________

Area	Total Number of Equipments	Number of Calibration Due	Number of Calibration		Number of Calibration		Any other Details
			On Time	Delayed	With Adjustment	Without Adjustment	
Production (Solid)							
Production (Liquid)							
Production (Sterile)							
QC labs							
R&D labs & pilot plant							
QA in-process (Solid)							
Biotech plant							
Cephalosporin & penicillin plants							
Weighing area (solid & liquid plants)							
Water treatment plants							
HVAC							
Stores							

Information provided by: _______________________

Name and signature: ___________________________

Date: _______________________________________

Key performance indicator: PREVENTIVE MAINTENANCE OF HEPA FILTERS AS PER SCHEDULE AND BREAK DOWN MAINTENANCE

Report for the month of _______________, year_________________

Scheduled Maintenance of HEPA Filters

Sterile Plant

Class	No. of Filters	Maintenance Plan/Due	Maintenance Performed	Observation
Class 100				
Class 1000				
Class 10,000				

Sterile Powder Plant

Class	No. of Filters	Maintenance Plan/Due	Maintenance Performed	Observation
Class 100				
Class 1000				
Class 10,000				

Biotechnology Plant

Class	No. of Filters	Maintenance Plan/Due	Maintenance Performed	Observation
Class 100				
Class 1000				
Class 10,000				

Information provided by: _______________________

Name and signature: ___________________________

Date: _______________________________________

49.4.1.5 Quality Control

Key performance indicator: TOTAL NUMBER OF RAW MATERIAL RECEIVED/RELEASED/PENDING

Report for the month of _______________, year_________________

Total Raw Material Received[1]	No. of Raw Material Released[2]	No. of Raw Material Pending[3]	Reason for Pending[4]

Information provided by: _______________________

Name and signature: ___________________________

Date: ______________________________________

Note:

1. Include total number of lots of raw material received in the lab, from the receiving system database.
2. In this column, provide information of the released number of materials.
3. How many of the total is pending.
4. Provide the reasons for pending, for example, lack of resources, under investigation for OOS results, least in priority, and so on.

Key performance indicator: OUT OF SPECIFICATION (OOS) INVESTIGATION RAW MATERIAL

Report for the month of ________________, year___________________

Total number of OOS: (RM) ______________________________________

Raw Material	Manufacturer/ Supplier	ABC Pharmaceutical Lot No.	Non-conformity	Investigation/ Corrective Action	Preventive Action

Information provided by: __________________________

Name and signature: ______________________________

Date: ____

Key performance indicator: TOTAL NUMBER OF BULK AND FINISHED PRODUCT RECEIVED/RELEASED/PENDING

Report for the month of ________________, year__________________

Plant	Total Bulk Product Received[1]	No. of Bulk Product Released[2]	No. of Bulk Product Pending[3]	Reason for Pending[4]
Solid dosage				
Liquid dosage				
Sterile				
Sterile powder				
Penicillins				
Cephalosporins				
Biotechnology plant				

Information provided by: ________________________

Name and signature: ___________________________

Date: ____

Note:

1. Include total number of lots of bulk product received in the lab, from the receiving system database.
2. In this column, provide information of the released number of bulk products.
3. Information about the total number which is pending.
4. Provide the reasons for pending, for example, lack of resources, under investigation for OOS results, least in priority, and so on.

Key performance indicator: OUT-OF-SPECIFICATION (OOS) INVESTIGATION BULK AND FINISHED PRODUCTS

Report for the month of _______________, year__________________

Total number of OOS: (BULK/FINISHED PRODUCTS) ___________________________

Product Name	Batch No.	Nonconformity[1]	Investigation/ Corrective Action[2]	Preventive Action[3]

Information provided by: ________________________

Name and signature: ____________________________

Date: ____

Note:

1. Mention the nature of nonconformance/characteristic test failing to meet specifications.
2. Mention if the investigations and corrective actions were taken.
3. What preventive measure is recommended?

Key performance indicator: TOTAL NUMBER OF PACKAGING COMMODITIES RECEIVED/RELEASED/PENDING

Report for the month of ________________, year__________________

Total No. of Packaging Material Received	No. of Packaging Material Released	No. of Packaging Material Pending	Reason for Pending

Information provided by: ________________________

Name and signature: ____________________________

Date: ____

Key performance indicator: OUT-OF-SPECIFICATION (OOS) INVESTIGATION PACKAGING MATERIAL

Report for the month of ________________, year__________________

Total number of OOS: (PM) ______________________________________

Commodity[1]	Supplier[2]	QC No.[3]	Nonconformity[4]	Investigation/ Corrective Action[5]	Preventive Action[6]

Information provided by: _________________________

Name and signature: _____________________________

Date: ____

Note:

1. Commodity's name for which nonconformance took place.
2. Name of supplier/manufacturer of the corresponding commodity.
3. Number assigned to the batch from QC lab.
4. Nature of nonconformance/failure.
5. Mention if investigations and corrective actions were carried out.
6. What preventive measure is recommended?

Key performance indicator: NONCONFORMANCE (REJECTED) PACKAGING MATERIAL

Report for the month of ______________, year________________

Total nonconformance: ______________________________ Nos.

Nonconformance use as is: ______________________________ Nos.

Total rejected: ______________________________ Nos.

Serial No.	Packaging Commodity Name[1]	QC No.[2]	Code No.[3]	Disposition[4]		Quantity[5]	Non-conformance Details[6]
				NC Use as is	NC Rejected		

Information provided by: ________________________

Name and signature: ____________________________

Date: __________   ________________

Note:

1. Name of the packaging commodity failing specifications.
2. Number assigned from QC lab.
3. Code number of the commodity.
4. Disposition: mention number of the materials for which the decision was taken to reject or to use on "As is basis."
5. How much was the quantity of material being rejected.
6. Details of nonconformance.

Key performance indicator: NONCONFORMANCE (REJECTED) RAW MATERIAL

Report for the month of _______________, year_________________

Total nonconformance: ______________________________ Nos.

Nonconformance use as is: ______________________________ Nos.

Total rejected: ______________________________ Nos.

Serial No.	Raw Material Name[1]	Lot No.[2]	Code No.[3]	Disposition[4]		Quantity[5]	Nonconformance Details[6]
				NC Use as is	NC Rejected		

Information provided by: ________________________

Name and signature: ____________________________

Date: ____

Note:

1. Name of the material failing specifications.
2. Number assigned to the lot from QC lab.
3. Code number of the material.
4. Disposition: mention number of the materials for which the decision was taken to reject or to use on "As is basis."
5. How much was the quantity of material being rejected.
6. Details of nonconformance.

Key performance indicator: NONCONFORMANCE (REJECTED) BULK MATERIAL

Report for the month of ________________, year__________________

Total nonconformance: ____ Nos.
Total rejected: ____ Nos.

Serial No.	Material Name[1]	Batch No.[2]	Code No.[3]	Disposition[4]		Quantity[5]	Nonconformance Details[6]
				NC Use as is	NC Rejected		

Information provided by: ________________________

Name and signature: ____________________________

Date: ____

Note:

1. Name of the bulk material failing specifications.
2. Number assigned to the batch from the QC lab.
3. Code number of the bulk material.
4. Disposition: mention number of the materials for which the decision was taken to reject or to use on "As is basis."
5. How much was the quantity of material being rejected.
6. Details of nonconformance.

Key performance indicator: NONCONFORMANCE (REJECTED) FINISHED PRODUCT

Report for the month of _______________, year_________________

Total nonconformance: _______________________________________ Nos.
Total rejected: _______________________________________ Nos.

Details of NCs Used As Is

Serial No.	Product Name	Batch No.	QC No.	NC (Rejected)	Quantity	Nonconformance Details

Information provided by: ______________________

Name and signature: __________________________

Date: _______________________________________

Key performance indicator: MICROBIOLOGICAL MONITORING OF PURIFIED WATER (SOLID, LIQUID, PENS & CEFS. AND WATER TREATMENT PLANTS)

Report for the month of ________________, year__________________

Total samples: ______________________________________
Number of nonconformance: ______________________________________
Over action limit (OAL): ______________________________________
Too numerous to count (TNTC): ______________________________________

Area	No. of Test[1] Performed	Sample OAL[2]	Sample TNTC[3]	Remarks[4]
Solid dosage				
Liquid dosage				
Pens. & Cefs.				
WTP				

Information provided by: ________________________

Name and signature: ___________________________

Date: ____

Note:

1. Total number of tests performed for each type of product in sterile and biotechnology areas.
2. Number of samples found with OAL.
3. Number of samples found with "Too numerous to count."
4. Remarks, if any.

Subjected Performance Indicator: MICROBIOLOGICAL MONITORING OF WATER FOR INJECTION (STERILE, BIOTECHNOLOGY, STERILE POWDER FILLING)

Report for the month of ______________________________, **year**_________________

Total number of samples: ____
Number of nonconformance: ____
Over action limit (OAL): ____
Too numerous to count (TNTC): ____

Area	No. of Test[1] Performed	Sample OAL[2]	Sample TNTC[3]	Remarks[4]
Sterile				
Sterile powder				
Biotechnology				
WFI plant				

Information provided by: ______________________

Name and signature: ______________________

Date: ____________________________________

Note:

1. Total number of tests performed for each type product in sterile and biotechnology areas.
2. Number of samples found with OAL.
3. Number of samples found with "Too numerous to count."
4. Remarks, if any.

Key performance indicator: CHEMICAL & PHYSICAL MONITORING OF PURIFIED WATER (SOLID, LIQUID, PENICILLINS, CEFALOSPORINS, AND WATER TREATMENT PLANTS)

Report for the month of ________________, year__________________

Total samples: ___________________________ Nos.

Locations	Activity	Frequency	No. of Test Performed	Results	Remarks

Information provided by: __________________________

Name and signature: ______________________________

Date: ____

Key performance indicator: VENDOR EVALUATION/VENDOR APPROVAL (SAMPLES RECEIVED) RAW/PACKAGING MATERIALS

Report for the month of _______________, year_________________

	No. of Samples Received	Approved	Rejected	Pending	Remarks
Raw material					
Packaging material					

Information provided by: _______________________

Name and signature: ___________________________

Date: _____________________________________

49.4.1.6 Packaging

Key performance indicator: REPACKING/SORTING

Report for the month of _______________, year_________________

Product	Batch No.	Rework/ Sorting[1]	Reason[2]	Corrective Action[3]	Preventive Measure[4]

Information provided by: _______________________

Name and signature: __________________________

Date: ____

Note:

1. Total number of batches involved in reworking or sorting.
2. Reason and root cause of rework.
3. What corrective action was taken?
4. What preventive measure was taken?

Key performance indicator: ABNORMAL VARIANCE (FINISHED PRODUCT & PACKAGING)

Report for the month of _______________, year__________________

Finished Product[1]	Batch No.[2]	Date	Standard Yield [3]	Actual[4]	Root Cause[5]	Corrective Action[6]	Preventive Action[7]

Information provided by: ________________________

Name and signature: ____________________________

Date: ____

Note:

1. Name of the finished product involved in the abnormal variance.
2. Batch number of the product.
3. Write down standard yield of the corresponding product as per master manufacturing formula.
4. What is the actual yield of the batch?
5. Root cause, if determined.
6. What corrective action was taken?
7. What preventive action was taken or recommended?

Key performance indicator: PROBLEMS DURING PACKAGING (COMMODITIES RELATED & OTHER BREAKDOWNS)

Report for the month of _______________, year__________________

Product[1]	Batch Number[2]	Code Number[3]	Commodity[4]	Nature of Problem[5]	Preventive Measure[6]	Closed Out[7]	
						Yes	No

Information provided by: ________________________

Name and signature: ___________________________

Date: ____

Note:

1. Name of the finished product involved in the problem during packaging.
2. Batch number of the product.
3. Code number of the product.
4. Commodity associated with the problem, for example, bottle, cartons, labels, leaflets, and so on.
5. Briefly add the nature of actual problem.
6. What preventive action was taken or recommended?
7. If the report closed out, mark Yes or No.

Key performance indicator: TOTAL NUMBER OF BATCH MANUFACTURED AND PACKED FOR EACH DOSAGE FORM

Report for the month of _______________, year_________________

Dosage Forms	Total No. of Batches	Total (units/numbers)
Tablets		
Vials (powder fill)		
Caplets		
Capsules		
Ampoules/vials (liquid fill)		
Suppositories		
PPS (granules)		
Spray/inhaler		
Cream		
Sterile cream		
Ointment		
Suspension		
Drops		
Vials (Lyophilized)		

Information provided by: _______________________

Name and signature: ___________________________

Date: _______________________________________

Key performance indicator: SIGNIFICANT EVENTS

Report for the month of ______________, year________________

1
2
3
4
5
6
7
8
9
10
11
12
13
14
15
16

Information provided by: ______________________

Name and signature: _________________________

Date: ____________________________________

Key performance indicator: FINAL BATCH DOCUMENTS AUDIT FINDINGS

Report for the month of _______________, year_________________

Dosage Form	No. of Batch Files Audited[1]	No. of Observations[2]	Corrective Actions Completed[2]	Pending Points[2]	Remarks[3]
Tablets/caplet					
Capsule					
Granules					
Syrup					
Suspension					
Drops					
Suppositories					
Spray/inhaler					
Sachets (liquid fill)					
Cream					
Ointment					
Ampoules/vials (liquid fill)					
Vials (powder fill)					
Sterile powder for injection					

Information provided by: ________________________

Name and signature: ____________________________

Date: ____

Note:

1. In this column total number of batch files audited for each dosage form will be entered.
2. Add here, if observations were made.
3. Number of observations for which corrective actions have been taken.
4. Number of observations for which corrective actions are pending.
5. Remarks, if any.

Key performance indicator: CLEANING VALIDATION

Report for the month of _______________, year_________________

Equipment/ Plant	No. of Protocol Finalized	No. of Samples Taken	No. of Analysis Performed in QC	No. of Samples Pending	No. of Reports Finalized	Remarks

Information provided by: ________________________

Name and signature: ____________________________

Date: ____

Key performance indicator: THE GMP/GLP/GOOD DOCUMENTATION TRAINING

Report for the month of ________________, year__________________

A. Existing local employees in house training

Department	No. of Employees Training Imparted to	GMP	GLP	Good Documentation	Other	Hours Spent	Remarks

B. New Employees

No. of New employees imparted induction training: _______________________

Information provided by: _______________________

Name and signature: ___________________________

Date: _______________________________________

Key performance indicator: THE GMP/GLP/GOOD DOCUMENTATION TRAINING

Report for the month of ________________, year__________________

Department	No. of Employees Training Imparted to	GMP	GLP	Good Documentation	Other	Hours Spent	Remarks

Information provided by: ________________________

Name and signature: ____________________________

Date: ____

Key performance indicator: RESEARCH & DEVELOPMENT LABORATORY ACTIVITIES (STABILITY/METHOD DEVELOPMENT/NEW PRODUCT)

Report for the month of ______________, year________________

Formulation Laboratory

Serial No.	Description	Total No.
1	Experimental batches	
2	Stability/pilot/batches	
3	Validation batches	
4	Master formula (new/updated)	
5	Problem resolution and formula optimization	

Analytical Laboratory

Serial No.	Description	Total No.
1	Standard test method	
2	Test method validation	
3	Comparative study (in vitro bioequivalence study)	
4	Sample analysis (registration & tender)	
5	Stability study (formal & ongoing)	
6	Problem resolution	

Packaging Laboratory

Serial No.	Description	Total No.
1	Artwork & text approval	
2	Initial sample for registration	
3	Final packs for registration approval (packaging material specification)	
4	Primary packaging material	

Documentation

Serial No.	Description	Total No.
1	Regulatory documents (development pharmaceutics, Mfg, flow chart)	
2	Updating specification	
3	Technical support for regulatory agency's queries	
4	Raw material specification, purchase specification	

Information provided by: ______________________

Name and signature: __________________________

Date: _____________________________________

Key performance indicator: REGULATORY AFFAIRS ACTIVITIES (SUBMISSION/ APPROVAL/MINOR CHANGES ETC.)

Report for the month of ________________, year__________________

Serial No.	Product Name	Date of Submission	Approved	Pending	Changes Requested	Remarks

Information provided by: ________________________

Name and signature: ____________________________

Date: ____

QAS-031

YOUR COMPANY STANDARD OPERATING PROCEDURE

SOP No. QAS-031.00 Effective date: mm/dd/yyyy
Approved by:

TITLE: **Vendors Evaluation and Approval Procedure**

AUTHOR: ______________________________

Name/Title/Department

____

Signature/Date

CHECKED BY: ______________________________

Name/Title/Department

____

Signature/Date

APPROVED BY: ______________________________

Name/Title/Department

____

Signature/Date

REVISIONS:

No.	Section	Pages	Initials/Date

SOP No. QAS-031.00 Effective date: mm/dd/yyyy
Approved by:

50.1 PURPOSE

To describe the procedure of vendor evaluation to supply the raw materials, packaging materials, and semifinished bulk of consistent quality conforming to the ABC Pharmaceutical Company's approved specifications.

50.2 RESPONSIBILITY

The QC manager, QA manager, and purchase manager are responsible for following the vendor approval system. The QC director and quality affairs director are responsible for the SOP compliance.

50.3 PROCEDURE

50.3.1 Responsibilities of Purchase Department

1. The purchase department will send purchasing specifications to the vendor who would commit his material to comply with the pharmacopeial- and/or Julphar-approved purchasing specification.
2. Based on the specifications, the purchase department shall provide samples at least from one lot and preferably from three lots to be evaluated by the QC department.
3. All samples should be submitted to the QC department with an analysis request form and a copy of it is sent to the QA manager (refer to Attachment I).
4. The documentation requirement for each type of material is specified in Table 50.1. It is the responsibility of the purchase department to provide all the documentation requirement specified in Table 50.1 to be accompanied with a vendor approval sample analysis requirement form at the time of sample submission with a copy to the designated QA manager.
5. It is the responsibility of the purchase manager to follow all the documents with the vendor to be responded along with the vendor approval samples.

50.3.2 Responsibilities of QC Department

1. All the samples received by the QC department shall be tested as per the pharmacopeial specifications on FIFO basis or as per the urgency of the material. All samples received will be entered in a designated register.
2. The sample quantity should be sufficient for at least three complete analysis and should be kept as a reference sample.
3. A unanimous decision may be taken by R&D, production, and QA directors to request for a larger sample of material passing analysis for product development experiments when needed.
4. The QA manager will create a documentation file related to each vendor to keep the records of testing.
5. Only those vendors meeting the quality level of the ABC Pharmaceutical Company specification shall be dealt with. The purchase department shall keep the list of vendors approved by QC/QA.
6. All vendor approval sample analysis shall be carried out on a workbook like all other routine samples of raw materials.

SOP No. QAS-031.00 Effective date: mm/dd/yyyy
Approved by:

TABLE 50.1
Required Documents for Approval of New Vendors

Documentation with Samples	Active	Excipient	Primary Packaging Material	Secondary Packaging Material
Vendor approval sample lots	3	1	1	1
Vendor's information (brochure)	✓	✓	✓	✓
Vendor specifications	✓	✓	✓	(1)
ABC Pharmaceutical Company specifications	✓	✓	✓	✓
GMP certificate of the manufacturing site	✓ (2)	(2) or (3)	(2) or (3)	N/A
Copy of drug master file	(2)	✓ (6)	(3)	N/A
Site master file	✓	✓	✓	—
Manufacturing site license	✓	✓	✓	✓
Relationship letter between manufacturing, supplier, and Julphar or as applicable	✓	✓	✓	✓
Letter of FDA approval	✓	✓	✓	N/A
Certificate of European suitability	✓	✓ (1)	✓ (2) (3)	N/A
Supply agreement contract	✓	✓	✓	(1)
Quality questionnaire (Attachment II)	✓	✓	✓	✓
Material safety data sheet	✓	✓	N/A	N/A
ISO 9001:2000	✓ (4)	✓ (4)	✓ (4)	✓ (4)
ISO 14001	✓ (4)	✓ (4)	✓ (4)	✓ (4)

Notes: (1) If available. (2) FDA approval or certificate of suitability. (3) GMP certificate from the MOH of the country of origin or FDA approval. (4) Preferable.

7. The vendors qualifying the ABC Pharmaceutical Company specifications will be notified to the purchase manager by the QC director with a copy to the QA manager.
8. The QA manager will follow up documents from the purchase department or, if required, directly from the manufacturer/supplier.
9. The material source approved by the QC and QA directors cannot be changed unless otherwise justified and approved by the R&D director in coordination with the regulatory affairs manager.
10. Each vendor approval/vendor rejection based on the sample analysis will be supported with comments from the QC manager or director.
11. The QA manager will directly communicate with the vendor to provide the missing documents.
12. If anomalies or deficiencies are observed, more information will be requested with corrective actions report. Once full compliance is received, the QC manager will update the master plan with the help of the QA manager.
13. In case of material received from brokers, the supplier will be asked to provide the details of the manufacturers, including the certificate of analysis issued by the actual manufacturer of the material.
14. In case the manufacturer is new, it will pass through the same procedure defined in the SOP for approval of new vendors, and will be requested to provide all the necessary documents mentioned in Table 50.1.

SOP No. QAS-031.00 Effective date: mm/dd/yyyy
Approved by:

15. The list of approved vendors will provide the names of the manufacturers. The complete address of the same could be obtained from the certificate of analysis provided by the corresponding manufacturers.

50.3.3 Responsibilities of QA Department

1. The QA manager will review the vendor's assessment form received from the vendors.
2. Once all the documents are received and the QA manager reviews to ensure that the documents are adequate, the vendor's assessment form will be signed by the QA manager and approved by the QA director.
3. The assessment questionnaire (Attachment II) response received from the vendor will also be reviewed by the QA manager and approved by the QA director to assure the vendor is in GMP compliance.

The list of vendors in the process of approval will be as follows:

List A: Vendors qualifying on the basis of first three successful lots, test history (0% rejection), and satisfactory document and GMP review
List B: Vendors qualifying a minimum of three lots, with rejection rate 0%, and documentation and GMP compliance review in progress
List C: Vendors qualifying a minimum of three lots, with rejection rate NMT 2%, and documentation and GMP compliance review in progress
List D: Vendors qualifying a minimum of three lots, with rejection rate NMT 5%, and documentation and GMP compliance review in progress

1. Vendor exceeding the rejection criteria of 5% will be communicated for corrective action or disqualified for a particular material permanently.
2. The QA manager will inform the QA director about critical vendors to be physically inspected.
3. The QA manager will update the vendor approval master plan on an ongoing basis.
4. The QA manager will prepare the action plan for the vendor approval qualification.

50.3.4 Responsibilities of QA Director

1. The QA director will verify critical vendors to be inspected as per the QC manager's recommendation.
2. The QA director will seek necessary approval from the top management for vendor inspection.

50.4 REASON FOR REVISION

First time issued for the ABC Pharmaceutical Company.

50.5 ATTACHMENT FORMS

ABC PHARMACEUTICAL COMPANY

LEVEL 3
SOP No.: QAS-031
Issued on:
Revision No.:
Initiator name:

ATTACHMENT I

Analysis Request Form

From : Purchase manager
To : [ ] Quality control manager
[ ] R&D lab director
[ ]
[ ]

Date: / /

Subject: Sample of ______

Manufacturer name: ______

Supplier name: ______

Documentation with Samples	Active		Excipient		Primary Packaging Material		Secondary Packaging Material	
	(R)	(S)	(R)	(S)	(R)	(S)	(R)	(S)
Vendor approval samples lots	3		1		1		1	
Vendor's information (brochure)	✓		✓		✓		✓	
Vendor specifications	✓		✓		✓		(1)	
Julphar specifications	✓		✓		✓		✓	
GMP certificate of the manufacturing site	✓ (2)		(2) or (3)		(2) or (3)		N/A	
Copy of drug master file	(2)		✓ (6)		(3)		N/A	
Site master file	✓		✓		✓		–	
Manufacturing site license	✓		✓		✓		✓	
Relationship letter between manufacturing, supplier, and Julphar or as applicable	✓		✓		✓		✓	

ABC PHARMACEUTICAL COMPANY

LEVEL 3
SOP No.: QAS-031
Issued on:
Revision No.:
Initiator name:

Letter of FDA approval	✓		✓		✓		N/A	
Certificate of European suitability	✓		✓ (1)		✓ (2) (3)		N/A	
Supply agreement contract	✓		✓		✓		✓	
Quality questionnaire	✓		✓		✓		✓	
Material safety data sheet	✓		✓		N/A		N/A	
ISO 9001:2000	✓ (4)		✓ (4)		✓ (4)		✓ (4)	
ISO 14001	✓ (4)		✓ (4)		✓ (4)		✓ (4)	

Please send us your report in due course.

Thanks,

Sincerely

Received by ______________________

Date ______________________

Purchase manager

1: If available
2: FDA approval or certificate of suitability
3: GMP certificate from the MOH of the country of origin or FDA approval
4: Preferable
S: Status
R: Required documents

Documents sent to the QC manager with the sample: mark ✓ (yes) or × (no)

ABC PHARMACEUTICAL COMPANY

LEVEL 3
SOP No.: QAS-031
Issued on:
Revision No.:
Initiator name:

ATTACHMENT II

Email address: ____________________

Dear **Mr.** ____________________

M/S ____________________

SUBJECT: **Vendor Approval Requirement**

You are requested to fill and return the **documentation** status checklist for each material **(Annexure A)** at your earliest to the undersigned to be forwarded to the ABC Pharmaceutical Company QA and regulatory affairs department.

The information is mandatory to raw material and semifinished bulk materials import to the company.

It is requested to kindly fill **Annexures B, C, D, and E** mentioned in **Annexure A** and send the related documents (originals) at your earliest to the undersigned.

Based on your feedback, the specific documents will be directly requested by our QA and RA departments as applicable.

Name
QA director
Email address:

Name
RA director
Email address:

Kindly provide the documents requested **(as per Annexure A)** for the following material and forward the same to Julphar purchase department at your earliest.

Names of the materials:

____

____

____

____

Regards,

ABC PHARMACEUTICAL COMPANY

ANNEXURE A

DOCUMENT STATUS CHECKLIST

Company name: ______________________________ Date: __________

Information provided by: ____________ Designation: __________ Signature: ____________

Serial No.	Documents for Registration with MOH UAE	Status*	Remarks
1.	a. **Legalized current GMP certificate issued by the competent authority in the country of origin (attested true)**		
	b. Legalized valid manufacturing license issued by the competent authority in the country of origin (attested true)		
	c. List of the products manufactured and/or assembled by the site		
	d. Site master file		
	e. A notarized copy of registration certificate or GMP evidence from other countries for the manufacturing site		
2.	**Declaration letter (for suppliers and brokers) for business association with manufacturer and the ABC Pharmaceutical Company. Draft copy attached (Annexure B)**		
Serial No.	**Other Supporting Documents for Registration with MOH (Mandatory)**	**Status***	**Remarks**
1.	*FDA approval letter*		
2.	*Certificate of European suitability*		
3.	Copy of drug master file (open part)		
4.	Quality questionnaire (self-evaluation questionnaire)—**Annexure C (for nonsterile); Annexure D (for sterile)**		
5.	Quality agreement, **Annexure E**		
6.	Material in-house specification		

Serial No.	Other Supporting Documents for Registration with MOH (Optional)	Status*	Remarks
1.	Company information/brochure (or website address)		
2.	Material safety data sheets for the APIs/excepients		
3.	ISO 9001-2000 certificate copy		
4.	ISO 14001-2004 certificate copy		

*Status: Mark
Available ✓
Not available ×

Remarks: Provide details if applicable or cannot provide.
Note: Annexures A, B, C, D, E to be filled and returned to the purchase manager, Julphar.

ANNEXURE B

Date: _______________

DIRECTOR OF DRUG CONTROL DEPARTMENT
MINISTRY OF HEALTH
(COUNTRY)

SUBJECT: Declaration and Quality Commitment

_______________________ (material name/s) _________________________

_______________________ (material name/s) _________________________

_______________________ (material name/s) _________________________

________________ (for ABC pharma product name/s) __________________

Dear sir,

We hereby certify that __________________________ (manufacturer name) ____________

(address): ____

Manufactures

and _______________________________ (supplier name) _____________________________

(address): ____

Supplies

______________ **(Raw material name/s)** _________________

____

____

To **ABC Pharmaceutical Company**—with a commitment that all applicable cGMP requirements and regulation will be followed and any changes to the process, materials, supplier, manufacturer, and so on will be informed as deemed necessary and in due course

Name and Title of Responsible Person
(Signature & Stamp)
For **M/S.** ______ **(Supplier)** __________

Name and Title of Responsible Person
(Signature & Stamp)
For **M/S.** ______ **(Manufacturer)** ________

ABC PHARMACEUTICAL COMPANY

ANNEXURE C (FOR NONSTERILE) MANUFACTURING FACILITY SELF-EVALUATION QUESTIONNAIRE

Kindly fill the following questionnaire and return to the ABC Pharmaceutical Company for review and comments to accomplish the vendor approval program.

Company Information:

Company name: ______________________________

Address: ______________________________

City: ______________ State: ______________

Zip: ______________ Country: ______________

Telephone (Plant): ______________ (Main Office): ______________

Name of information provider: ______________________________

Designation: ______________________________

Please provide

a. Organization chart.
b. Copy of ISO 9001:2000 and/or other certificate bodies along with the filled questionnaire.

ABC PHARMACEUTICAL COMPANY

ANNEXURE C (FOR NONSTERILE)
MANUFACTURING FACILITY
SELF-EVALUATION QUESTIONNAIRE

How is a lot/batch defined?

____

____

What is the lot number system?

____

Approximate number of employees at the facility? ____________________

Approximate square footage of facility (including warehouse)? ____________________

Management profile:

Facility plant manager: ____________________

Quality control manager: ____________________

Production manager: ____________________

Has the FDA or DEA inspected the facility? If yes, provide dates and results.

____

____

1st Quotation/Bid System and Customer Service	**Yes**	**No**	**N/A**
1. Are customer specifications reviewed by QC and/or groups to assure the requirements can be satisfied?	☐	☐	☐
2. Is there a system to generate reliable price quotes and return them by the due date?	☐	☐	☐
3. Are PO due dates verified against a production schedule before confirmation to the customer?	☐	☐	☐
4. Can the vendor provide the status of customer orders in process?	☐	☐	☐
5. Is the system of control adequate to assure that current and complete specifications are on file and will be utilized in manufacturing the item?	☐	☐	☐
6. Is there a system in place to notify customers of changes in manufacturing site, lead time, specification, process, or acceptance criteria?	☐	☐	☐
7. Is there a system to handle customer complaints and respond in a timely manner?	☐	☐	☐
8. Does the firm have current and applicable SOP on supplier audit and internal audit?	☐	☐	☐

ABC PHARMACEUTICAL COMPANY

ANNEXURE C (FOR NONSTERILE) MANUFACTURING FACILITY SELF-EVALUATION QUESTIONNAIRE

Comments: ______________________________

____

____

____

____

____

____

____

Information provided by: ______________________________

Designation: ______________________________

2nd Receiving Control	**Yes**	**No**	**N/A**
1. Are released items properly segregated from the ones awaiting testing and disposition?	☐	☐	☐
2. Is there an acceptable area and approved procedure for sampling?	☐	☐	☐
3. Is rejected material adequately controlled?	☐	☐	☐
4. Are material stored properly to prevent damage and mix-up?	☐	☐	☐
5. Where required, is the controlled environment storage adequate (humidity, temperature, etc.)?	☐	☐	☐

Comments: ______________________________

____

____

____

Information provided by: ______________________________

Designation: ______________________________

ABC PHARMACEUTICAL COMPANY

ANNEXURE C (FOR NONSTERILE) MANUFACTURING FACILITY SELF-EVALUATION QUESTIONNAIRE

3rd Manufacturing Facilities, Equipment, and Control	Yes	No	N/A
1. Are lubricants and nonproduction materials properly controlled to prevent contamination?	☐	☐	☐
2. Are the manufacturing procedures formalized, controlled, and validated?	☐	☐	☐
3. Are current specifications readily accessible for the employees?	☐	☐	☐
4. Are production areas that are very dusty or unique in regard to contamination properly controlled with exhaust systems or other methods of decontamination?	☐	☐	☐
5. Are any special requirements for specific operator apparel (masks, gowns, etc.) being violated?	☐	☐	☐
6. Are waste materials clearly identified in proper containers?	☐	☐	☐
7. Are the ceiling fixtures and pipes free of accumulated dirt, dust, and paint chips to prevent contamination of packaging materials?	☐	☐	☐

Comments: ____

____

____

____

Information provided by: ______________________

Designation: ______________________

ABC PHARMACEUTICAL COMPANY

ANNEXURE C (FOR NONSTERILE) MANUFACTURING FACILITY SELF-EVALUATION QUESTIONNAIRE

4th Quality Control	Yes	No	N/A
1. Is USP organic volatiles and other residual solvents established for drug substance?	☐	☐	☐
2. Does firm have a SOP for complaints handling, follow-up, documentation, and preventive action?	☐	☐	☐
3. Are all raw data properly recorded and available for verification?	☐	☐	☐
4. Does the firm have SOP for handling abnormal data?	☐	☐	☐
5. Is the firm capable of performing microbial testing and endotoxin test for packaging materials?	☐	☐	☐

Comments: ______________________________

____

____

____

Information provided by: ______________________________

Designation: ______________________________

5th Documentation Control	Yes	No	N/A
1. Does the facility have an adequate system to control specifications, test methods, and other documents?	☐	☐	☐
2. Are records maintained that identify the reasons for changes in documents?	☐	☐	☐
3. Are production records retained for adequate length of time?	☐	☐	☐
4. Are employee training records maintained for adequate length of time?	☐	☐	☐

Comments: ______________________________

____

____

____

Information provided by: ______________________________

Designation: ______________________________

ABC PHARMACEUTICAL COMPANY

ANNEXURE C (FOR NONSTERILE) MANUFACTURING FACILITY SELF-EVALUATION QUESTIONNAIRE

6th Calibration Control (Manufacturing/Laboratory)	**Yes**	**No**	**N/A**
1. Are there acceptable procedures to assure that gauges, tools, and test equipment are maintained in proper order?	☐	☐	☐
2. If the calibration is performed by the facility, are there calibration procedures formalized?	☐	☐	☐
3. Is there a system to assure that gauges and testing equipment are calibrated on time by responsible personnel?	☐	☐	☐
4. If computer software is used for scheduling preventative maintenance and calibrations, is it validated?	☐	☐	☐

Comments: ____

____

____

____

Information provided by: ____________________

Designation: ____________________

7th Packaging/ Shipping	**Yes**	**No**	**N/A**
1. Are the shipping containers sufficiently marked to identify content, storage condition, lot number, country of origin, and quantity?	☐	☐	☐
2. Are shipments adequately protected from in-transit damage?	☐	☐	☐
3. If drums are recycled from clients, are they cleaned via a validated procedure?	☐	☐	☐
4. Does the company include a certificate of analysis with the shipment?	☐	☐	☐
5. Does the company include a packing list with the shipment?	☐	☐	☐

Comments: ____

____

____

____

Information provided by: ____________________

Designation: ____________________

ABC PHARMACEUTICAL COMPANY

ANNEXURE C (FOR NONSTERILE) MANUFACTURING FACILITY SELF-EVALUATION QUESTIONNAIRE

8th Facility in General	**Yes**	**No**	**N/A**
1. Do adequate environmental controls exist in the production, warehouse, inspection, and laboratory areas?	☐	☐	☐
2. Are appropriate bathroom facilities available?	☐	☐	☐
3. Are unnecessary jewelry and cosmetics kept out of the production area?	☐	☐	☐
4. Is smoking and eating prohibited where necessary?	☐	☐	☐
5. Is there sufficient space for production operators, material handling, and staging?	☐	☐	☐
6. Is there an adequate pest control program in effect?	☐	☐	☐

Comments: ____

____

____

____

Information provided by: ____________________

Designation: ____________________

ABC PHARMACEUTICAL COMPANY

ANNEXURE C (FOR NONSTERILE) MANUFACTURING FACILITY SELF-EVALUATION QUESTIONNAIRE

Review Form (To be filled by ABC Pharmaceutical Company)

Vendor-related information (attach list wherever required)

1. **Items currently procured from source:**

   a. ____________ d. ____________

   b. ____________ e. ____________

   c. ____________ f. ____________

2. **Items intended to be procured from source:**

   a. ____________

   b. ____________

   c. ____________

3. **Are the materials under consideration for purchase produced only at this location?**

   Yes: ____________ No: ____________

   If not, where else do they originate from?

   ____

   ____

   ____

   ____

4. **Review of information provided by vendors (pages 1–8)**

   ____

   ____

   ____

   ____

   ____

5. **Conclusion:** ____

Reviewed by: ____________ Approved by: ____________

Date: ____________ Date: ____________

ABC PHARMACEUTICAL COMPANY

ANNEXURE D (FOR STERILE)
STERILE BULK MANUFACTURING FACILITY
SELF-EVALUATION QUESTIONNAIRE

Kindly fill the following questionnaire and return to the ABC Pharmaceutical Company's QA for review and comments to accomplish the vendor approval program.

Company Information:

Company name: ______________________________

Address: ______________________________

City: ______________ State: ______________

Zip: ______________ Country: ______________

Telephone (Plant): ______________ (Main Office): ______________

Name of information provider: ______________________________

Designation: ______________________________

Please provide

a. Organization chart.

b. Copy of ISO 9001:2000 and/or other certificate bodies along with the filled questionnaire.

ABC PHARMACEUTICAL COMPANY

ANNEXURE D (FOR STERILE)
STERILE BULK MANUFACTURING FACILITY
SELF-EVALUATION QUESTIONNAIRE

How is a lot/batch defined?

____

____

What is the lot number system?

____

Is there a way to denote the site of manufacture in the lot numbering system if multiple sites are employed?

____

Approximate number of employees at facility ______________________

Approximate square footage of facility (including warehouse) ______________________

Management Profile:

Facility plant manager: ______________________
Quality control manager: ______________________
Production manager: ______________________

Has the FDA or DEA inspected the facility? If yes, provide dates and results.

____

____

DMF (or AADA) for drug substance:

Original file number: ______________ Date: ______________
Last DMF revision Date: ______________ Reason: ______________

____

A. Strengths
 1. ______________________
 2. ______________________
 3. ______________________
 4. ______________________
 5. ______________________

B. Weaknesses
 1. ______________________
 2. ______________________
 3. ______________________
 4. ______________________
 5. ______________________

ABC PHARMACEUTICAL COMPANY

ANNEXURE D (FOR STERILE)
STERILE BULK MANUFACTURING FACILITY
SELF-EVALUATION QUESTIONNAIRE

1st **Quotation/Bid System and Customer Service**	**Yes**	**No**	**N/A**
1. Are customer specifications reviewed by quality control and/or groups to assure the requirements can be satisfied?	☐	☐	☐
2. Is there a system to generate reliable price quotes and return them by the due date?	☐	☐	☐
3. Are purchase order due dates verified against a production schedule before the company sends material to the customer?	☐	☐	☐
4. Is there an adequate long-term capacity plan for the drug substance?	☐	☐	☐
5. Can the vendor provide the status of customer orders in process?	☐	☐	☐
6. Is the system of control adequate to assure that current and complete specifications are on file and will be utilized in manufacturing the item?	☐	☐	☐
7. Is there a system in place to notify customers of changes in manufacturing site, lead time, specification, process, or acceptance criteria?	☐	☐	☐
8. Is there a system to handle customer complaints and respond in a timely manner?	☐	☐	☐
9. Does the company have current and applicable SOPs on supplier audit and internal audit?	☐	☐	☐

Comments: ____

____

____

____

____

____

____

____

Information provided by: ____________________

Designation: ____________________

ABC PHARMACEUTICAL COMPANY

ANNEXURE D (FOR STERILE)
STERILE BULK MANUFACTURING FACILITY
SELF-EVALUATION QUESTIONNAIRE

2nd	Receiving Control	Yes	No	N/A
1.	Does the receiving inspection check incoming shipments to the requirement of the purchase order, specification, and applicable drawings?	☐	☐	☐
2.	Is there a list of qualified suppliers?	☐	☐	☐
3.	If not, is there a procedure to prevent acceptance of material from unqualified suppliers?	☐	☐	☐
4.	Are released items properly segregated from material awaiting testing and disposition?	☐	☐	☐
5.	If multiple shipment dates of the same lot or multiple lots in one shipment of the same item is received, is each lot received, sampled, tested, and released independently?	☐	☐	☐
6.	Is there an acceptable area and approved procedure for sampling?	☐	☐	☐
7.	Do material container and receiving inspection records indicate acceptance or rejection of incoming material?	☐	☐	☐
8.	Is the rejected material adequately controlled?	☐	☐	☐
9.	Are material stored properly to prevent damage and mix-up?	☐	☐	☐
10.	Are materials properly identified as to their contents to avoid mix-up in issuance?	☐	☐	☐
11.	Is there a first-in first-out (FIFO) system for ingredients?	☐	☐	☐
12.	Are records kept that track the date of receipt, lot number, and vendor?	☐	☐	☐
13.	Are stocks reinspected and tested at intervals?	☐	☐	☐
14.	Where required, is the controlled environment storage adequate (humidity, temperature, etc.)?	☐	☐	☐
15.	If receiving, release, issuance, and reject status controlled by computer is software-validated?	☐	☐	☐

Comments: ____

____

____

____

Information provided by: ____________________

Designation: ____________________

ABC PHARMACEUTICAL COMPANY

ANNEXURE D (FOR STERILE)
STERILE BULK MANUFACTURING FACILITY
SELF-EVALUATION QUESTIONNAIRE

3rd Manufacturing Facilities, Equipment, and Control	**Yes**	**No**	**N/A**
1. Are the items being manufactured identifiable throughout the operation?	☐	☐	☐
2. Are pipes labeled/color coded to show contents and flow direction?	☐	☐	☐
3. Are different grades of the same chemical distinguishable in manufacturing to prevent mix-up?	☐	☐	☐
4. Are the manufacturing operations orderly in flow and organization?	☐	☐	☐
5. Are the manufacturing processes and facilities adequate to produce quality products?	☐	☐	☐
6. Are lubricants and nonproduction materials properly controlled to prevent product contamination?	☐	☐	☐
7. If drums are recycled from clients, are they cleaned via validated cleaning procedure, visually inspected, and material tested before use?	☐	☐	☐
8. Are manufacturing equipment and utensils properly cleaned with a validated process between usages?	☐	☐	☐
9. Does the lot number reflect one homogeneous production run?	☐	☐	☐
10. Are the manufacturing procedures formalized, controlled, and validated?	☐	☐	☐
11. Is there a formal procedure for deviations to normal practice?	☐	☐	☐
12. Are there formal validated procedures for the handling of rework?	☐	☐	☐
Is it submitted in DMF?	☐	☐	☐
13. Are current specifications readily accessible for the employees?	☐	☐	☐
14. Are production areas that are very dusty or unique in regard to contamination properly controlled with exhaust systems or other methods of decontamination?	☐	☐	☐
15. Are any special requirements for specific operator apparel (masks, gowns, etc.) being violated?	☐	☐	☐

Comments: ____

____

____

____

Information provided by: ______________________

Designation: ______________________

ABC PHARMACEUTICAL COMPANY

ANNEXURE D (FOR STERILE)
STERILE BULK MANUFACTURING FACILITY
SELF-EVALUATION QUESTIONNAIRE

3rd Manufacturing Facilities, Equipment, and Control (continued)	**Yes**	**No**	**N/A**
16. Is there a formal record of production steps and controlled parameters (temperature, pH, time, etc.) and quantities used in the process?	☐	☐	☐
17. Are waste materials clearly identified in proper containers?	☐	☐	☐
18. Are the ceiling fixtures and pipes free of accumulated dirt, dust, and paint chips to prevent product contamination?	☐	☐	☐
19. Are valves and associated pipes in the production areas free of leaks?	☐	☐	☐
20. Is there a system to verify the use of only released components?	☐	☐	☐
21. If water is used in the process, does it meet compendia requirement for the type of water used?	☐	☐	☐
22. Are manufacturing equipment adequate in size and designed to produce maximum batch size quality product?	☐	☐	☐
23. Is the manufacturing process controlled by computer software-validated?	☐	☐	☐

Comments: ____

____

____

____

Information provided by: ____________________

Designation: ____________________

ABC PHARMACEUTICAL COMPANY

ANNEXURE D (FOR STERILE)
STERILE BULK MANUFACTURING FACILITY
SELF-EVALUATION QUESTIONNAIRE

4th Quality Control	**Yes**	**No**	**N/A**
1. Is the impurity profile of the drug substance established and individual impurities identified with validated methods above 0.1% level?	☐	☐	☐
2. Is USP organic volatiles and other residual solvents established for drug substance?	☐	☐	☐
3. Does the company have an annual product review SOP?	☐	☐	☐
4. Is there a formal stability program for drug substance?	☐	☐	☐
5. Are all stability chambers and incubators controlled for specified humidity and temperature requirements?	☐	☐	☐
6. Does the company have a SOP for complaints handling, follow-up, documentation, and preventive action?	☐	☐	☐
7. Are all raw data properly recorded and available for verification?	☐	☐	☐
8. Does the company have SOP for handling aberrant data?	☐	☐	☐
9. Is the company capable of performing microbial testing and endotoxin test for bulk drug substance?	☐	☐	☐

Comments: ____

____

____

____

Information provided by: ____________________

Designation: ____________________

ABC PHARMACEUTICAL COMPANY

ANNEXURE D (FOR STERILE)
STERILE BULK MANUFACTURING FACILITY
SELF-EVALUATION QUESTIONNAIRE

5th Documentation Control	**Yes**	**No**	**N/A**
1. Does the facility have an adequate system to control specifications, test methods, and other documents?	☐	☐	☐
2. Is there a provision to notify regulatory agencies, clients, and appropriate disciplines of changes in specification, process, and test methods?	☐	☐	☐
3. Are records maintained that identify the reasons for changes in documents?	☐	☐	☐
4. Are product development and laboratory records easily available?	☐	☐	☐
5. Are production records retained for adequate length of time?	☐	☐	☐
6. Are employee training records maintained for adequate length of time?	☐	☐	☐

Comments: ______________________________

____

____

____

Information provided by: ______________________

Designation: ______________________

ABC PHARMACEUTICAL COMPANY

ANNEXURE D (FOR STERILE)
STERILE BULK MANUFACTURING FACILITY
SELF-EVALUATION QUESTIONNAIRE

6th Calibration Control (Manufacturing/Laboratory)	**Yes**	**No**	**N/A**
1. Are there acceptable procedures to assure that gauges, tools, and test equipment are maintained in proper order?	☐	☐	☐
2. If calibration is performed by the facility, are there calibration procedures formalized?	☐	☐	☐
3. Is there a system to assure that gauges and testing equipment are calibrated on time by responsible personnel?	☐	☐	☐
4. If computer software is used for scheduling preventative maintenance and calibrations, is it validated?	☐	☐	☐

Comments: ____

____

____

____

Information provided by: ______________________

Designation: ______________________

ABC PHARMACEUTICAL COMPANY

ANNEXURE D (FOR STERILE)
STERILE BULK MANUFACTURING FACILITY
SELF-EVALUATION QUESTIONNAIRE

7th Packaging/Shipping	**Yes**	**No**	**N/A**
1. Are the shipping containers sufficiently marked to identify content, storage condition, lot number, country of origin, quantity, and expiration date, if applicable?	☐	☐	☐
2. Are expiration dates based on current stability data?	☐	☐	☐
3. Are shipments adequately protected from in-transit damage?	☐	☐	☐
4. If drums are recycled from clients, are they cleaned via a validated procedure?	☐	☐	☐
5. Is there a formalized procedure to inspect and clean by validated method for drums before filling for shipment?	☐	☐	☐
6. Are storage tanks dedicated for specific chemicals?	☐	☐	☐
7. Does the company include a certificate of analysis with the shipment?	☐	☐	☐
8. Does the company include a packing list with the shipment?	☐	☐	☐

Comments: ____

____

____

____

Information provided by: ______________________

Designation: ______________________

ABC PHARMACEUTICAL COMPANY

ANNEXURE D (FOR STERILE)
STERILE BULK MANUFACTURING FACILITY
SELF-EVALUATION QUESTIONNAIRE

8th Facility in General	**Yes**	**No**	**N/A**
1. Do adequate environmental controls exist in the production, warehouse, inspection, and laboratory areas?	☐	☐	☐
2. Are appropriate bathroom facilities available?	☐	☐	☐
3. Are unnecessary jewelry and cosmetics kept out of the production area?	☐	☐	☐
4. Is smoking and eating prohibited where necessary?	☐	☐	☐
5. Is there sufficient space for production operators, material handling, and staging?	☐	☐	☐
6. Is there an adequate pest control program in effect?	☐	☐	☐

Comments: ______________________________

____

____

____

Information provided by: ____________________

Designation: ____________________

ABC PHARMACEUTICAL COMPANY

ANNEXURE D (FOR STERILE)
STERILE BULK MANUFACTURING FACILITY
SELF-EVALUATION QUESTIONNAIRE

9th	**Sterile Bulk Manufacturer**	**Yes**	**No**	**N/A**
1.	Does the company have flowcharts with the major processing steps (including impurity)?	☐	☐	☐
2.	Does the company have piping system diagram for all utilities (e.g., WFI, clean steam, gases, etc.)?	☐	☐	☐
3.	Does the company have product validation reports for drug substances (all parts, process variables, phases, steps, fluid pathway, time limitation study)?	☐	☐	☐
4.	Are aseptic operations performed by operator?	☐	☐	☐
5.	Is culture media or nonmedia an alternative used to assure assessment of sterility for the aseptic process validation?	☐	☐	☐
	a. Is validation close to current practices?	☐	☐	☐
6.	Is environmental and personnel monitoring performed during validation to establish limits?	☐	☐	☐
7.	Are aseptic operations performed in a closed system?	☐	☐	☐
8.	Does the company monitor bioburden of nonsterile critical components and of operational steps?	☐	☐	☐
9.	Does the company have endotoxin removal validation report for the manufacturing process?	☐	☐	☐
10.	If yes, does it contain endotoxin level monitoring of starting components?	☐	☐	☐
11.	Is endotoxin removal validation for conversion from nonsterile to sterile performed with			
	a. Laboratory/pilot-scale batches	☐	☐	☐
	b. Production batches	☐	☐	☐
12.	Does aseptic processing facility include temperature, humidity, and pressure differential control for different class area?	☐	☐	☐
13.	Does the company have routine environmental monitoring program?	☐	☐	☐

Comments: ____

____

____

____

Information provided by: ____________________

Designation: ____________________

ABC PHARMACEUTICAL COMPANY

ANNEXURE D (FOR STERILE)
STERILE BULK MANUFACTURING FACILITY
SELF-EVALUATION QUESTIONNAIRE

9th Sterile Bulk Manufacturer (continued)	**Yes**	**No**	**N/A**
14. Does the company have a sanitization qualification program?	☐	☐	☐
15. Does the company have filter selection criteria and filtration validation program?	☐	☐	☐
16. Does the company have completed filter validation report for sterilizing filter with the drug substance?	☐	☐	☐
17. Does the company perform integrity testing of sterilizing filters?	☐	☐	☐
18. Are sterilizing filters changed after completion of each production lot?	☐	☐	☐
19. If filters(s) are not changed, does the the company have justification for intergrity of the filter for the time periods utilized (multiple batch usage) and data to show "grow through" has not occurred?	☐	☐	☐
20. Is the equipment used in processing of sterile bulk drug phase sterilized with validated process?	☐	☐	☐
21. Is the sanitization process of equipment, transfer lines performed rather than steam sterilization?	☐	☐	☐
22. Has the company determine the low spots or does it monitor at discharge during validation?	☐	☐	☐
23. Does the company have equipment maintenance log for sterilizing system?	☐	☐	☐
24. Does the company use rodac or surface plates for determination of level of contamination?	☐	☐	☐
25. Does the company have procedure for sampling of critical sufaces such as operator's gloves, surface of equipment?	☐	☐	☐
26. If reprocessing steps are performed, is it validated and included in DMF?	☐	☐	☐
27. Does the company use WFI in the final critical processing steps?	☐	☐	☐
28. Is the WFI system validated?	☐	☐	☐
29. Is the WFI monitored daily against compendial requirement?	☐	☐	☐

Comments: ______________________________

____

____

____

Information provided by: ______________________

Designation: ______________________

ABC PHARMACEUTICAL COMPANY

ANNEXURE D (FOR STERILE)
STERILE BULK MANUFACTURING FACILITY
SELF-EVALUATION QUESTIONNAIRE

9th Sterile Bulk Manufacturer (continued)	**Yes**	**No**	**N/A**
30. Does the the company have seal integrity validation data for container/closure?	☐	☐	☐
31. Does the the company have cleaning neutralization process validation?	☐	☐	☐
32. Does the the company have stability of bulk substance in the final storage container?	☐	☐	☐
33. Does the company have failure investigation process and logic for release for a batch manufactured in campaign?	☐	☐	☐
34. Does the company have procedure for revalidation?	☐	☐	☐
35. Is sterility test performed in-house?	☐	☐	☐
36. Does the company have qualification requirements of personnel responsible for sterility testing?	☐	☐	☐
37. Does the company have procedure for evaluating (releasing/rejecting) batches that fail the initial sterility tests?	☐	☐	☐
38. Does the company perform trend analysis for microbiological monitoring in central processing area?	☐	☐	☐
39. Are all noncompendial test methods validated?	☐	☐	☐

Comments: ____

____

____

____

Information provided by: ______________________

Designation: ______________________

ABC PHARMACEUTICAL COMPANY

ANNEXURE D (FOR STERILE)
STERILE BULK MANUFACTURING FACILITY
SELF-EVALUATION QUESTIONNAIRE

Review Form (To Be Filled by the ABC Pharmaceutical Company)

Vendor-related information (attach list wherever required)

1. **Items currently procured from source:**

   a. ____________ d. ____________

   b. ____________ e. ____________

   c. ____________ f. ____________

2. **Items intended to be procured from source:**

   a. ____________

   b. ____________

   c. ____________

3. **Are the materials under consideration for purchase produced only at this location?**

   Yes: ____________ No: ____________

   If not, where else do they originate?

   ____

   ____

   ____

   ____

4. **Review of information provided by vendors (pages 1–14):**

   ____

   ____

   ____

   ____

   ____

5. **Conclusion:** ____

Reviewed by: ____________ Approved by: ____________

Date: ____________ Date: ____________

ANNEXURE E

QUALITY AGREEMENT

This quality agreement is by and between M/S <manufacturer name> with office at ____________________ hereafter referred to as supplier and M/S ABC Pharmaceutical Company with office at <address of the company> hereafter referred to as ABC Pharmaceutical Company, whereas manufacturer supplies APIs/excipients suitable for pharmaceutical use for ABC Pharmaceutical Company.

This agreement pertains to the following primary APIs/excipients:

____

____

____

____

1. The manufacturer shall manufacture, test, and release the APIs/excipients in accordance with and will comply to
   a. WHO GMP guidelines and/or guidelines applicable to the countries in which the APIs/excipients is being registered, (e.g. FDA, EU, etc).
   b. Guidelines applicable to the country of origin and other regional certifications (if applicable).
   c. ISO-9001:2000, current version (optional).
   d. ISO-14001:2004, current version (optional).
2. This quality agreement will outline the responsibilities of the supplier and ABC Pharmaceutical Company with regard to the quality activities discussed in the quality criteria listed above.
3. The manufacturer/supplier shall not allow a third party to manufacture, package, label, or release the APIs/excipients unless agreed upon in writing with ABC Pharmaceutical Company.
4. M/S <Manufacturer> will be responsible to ensure that the specifications for APIs/excipients are in compliance with those sent by ABC Pharmaceutical Company.
5. Upon request by ABC Pharmaceutical Company, M/S <Manufacturer> shall disclose to ABC Pharmaceutical Company recent regulatory agency inspections and findings pertaining to the APIs/excipients.
6. <Manufacturer> shall notify promptly if in the course of a regulatory inspection, negative findings are made related to the quality of material supplied.
7. <Manufacturer> shall provide certificate of analysis with each lot of the corresponding APIs/excipients supplied.
8. <Manufacturer> shall communicate the recommended storage and transportation conditions of the APIs/excipients.
9. <Manufacturer> shall ensure that APIs/excipients are stored and supplied in accordance with recommended storage conditions.
10. <Manufacturer> have a written procedure to investigate and document quality-related complaints. A root cause analysis, action taken for correction of the problem, and the formal conclusion will be provided to the ABC Pharmaceutical Company within a reasonable time after receipt of the complaint related to APIs/excipients.
11. The parties shall cooperate in the exchange of information required to effectively conduct an investigation.

12. ABC Pharmaceutical Company has the right to audit <Manufacturer> facilities and system as they relate to the manufacturer of APIs/excipients at mutually agreed upon time or at least once per year.
13. The <Manufacturer> will keep ABC Pharmaceutical Company informed about the changes which might have an effect on the material's quality.

MANUFACTURER

Company name : ____________________
Designation : ____________________
Signature : ____________________
Date : ____________________

ABC Pharmaceutical Company

Designation : **Director quality assurance**
Signature : ____________________
Date : ____________________

QAS-032

YOUR COMPANY STANDARD OPERATING PROCEDURE

SOP No. QAS-032.00 Effective date: mm/dd/yyyy
Approved by:

TITLE: **Process Simulation (Media Fill) Test**

AUTHOR: ______________________________

Name/Title/Department

____

Signature/Date

CHECKED BY: ______________________________

Name/Title/Department

____

Signature/Date

APPROVED BY: ______________________________

Name/Title/Department

____

Signature/Date

REVISIONS:

No.	Section	Pages	Initials/Date

SOP No. QAS-032.00 Effective date: mm/dd/yyyy
Approved by:

Subject: Process Simulation (Media Fill) Test

51.1 PURPOSE

This SOP describes methods and procedures for conducting the process simulation test of sterile products, including media fill procedures, media selection, incubation time and temperature, inspection of filled vials/ampoules, documentation, interpretation of results, and possible corrective actions required. The main purposes of the SOP are to

- Demonstrate control over the aseptic operations
- Qualify aseptic processing (personnel, facility, equipment, environment, and procedures)
- Qualify aseptic techniques
- Comply with cGMP requirements

51.2 RESPONSIBILITY

It is the responsibility of all concerned department managers to follow the procedure. The manufacturing director, quality control director, and QA director are responsible for SOP compliance.

1. The manufacturing departments will be responsible for the following:
   - 1.1 Scheduling of the validation runs
   - 1.2 Raise media fill request form to all concerned departments with schedule (Attachment No. I)
   - 1.3 Provide operators involved in process simulation
   - 1.4 Indicating responsible supervisory personnel
   - 1.5 Perform media fill as per MFM and approved protocols
   - 1.6 Provide the material and utilities to be used
   - 1.7 Preparation of media
   - 1.8 Clean the line after the run and prepare the area for real work
   - 1.9 Review any major repairs
   - 1.10 Review calibration status
2. The engineering and maintenance departments are responsible for the following:
   - 2.1 Calibration of process equipment instrumentation on a regularly scheduled basis and/or after repairs
   - 2.2 Completion of major repairs/renovations prior to validation runs, if any
   - 2.3 Provide historical data for temperature, humidity, pressure, and so on
   - 2.4 Provide historical data for WFI and DI used during the media fill process
   - 2.5 Participation and/or presence during media fill qualification
3. Quality control is responsible for the following:
   - 3.1 Testing the microbial contamination and inspection of vials/ampoules
   - 3.2 Area monitoring for viable count
   - 3.3 Water testing
   - 3.4 Media testing (bulk media and filled vials)
   - 3.5 Compressed air testing (microbiologically)
   - 3.6 Nitrogen gas testing (microbiologically)
   - 3.7 CIP water (if applicable)

SOP No. QAS-032.00 Effective date: mm/dd/yyyy
Approved by:

4. Quality assurance is responsible for
   4.1 Process inspection/monitoring
   4.2 Air monitoring for nonviable counts
   4.3 In-process sampling (if required)
   4.4 Batch documents review
6. The validation group is responsible for the following:
   6.1 MFM initiation and review with the concerned departments
   6.2 Protocols initiation and review with the concerned departments
   6.3 Monitoring of completeness, accuracy, technical excellence, and applicability
   6.4 Scheduling of the validation runs (in conjugation with the manufacturing department)
   6.5 Conducting of the validation runs including recording of all data, and so on
   6.6 Data review and validation run acceptance
   6.7 Final report preparation
   6.8 Scheduled revalidation
7. Assistant manager QA, microbiological section head, QC director, sterile area production manager, operations manager, manufacturing director, and QA director will evaluate and approve the results in case of failure and determine the remedial actions.

51.3 FREQUENCY

51.3.1 Initial Performance Qualification

1. Three media fill runs are required for the following (this list is not all inclusive):
   1.1 New filling room or machine
   1.2 Major room modification (such as wall or ceiling reconfigurations)
   1.3 New equipment or new machine, which has product contact or affects product flow, which is not an exact equivalent of the original equipment
   1.4 Major mechanical or line configuration changes (such as new equipment additions or major preventative maintenance)
   1.5 Major HVAC changes, which may impact air flow patterns related to the critical filling area (Class 100) (such as addition/removal of HEPA housings)

51.3.2 Periodic Performance Requalification

At least every 6 months for each product/container configuration and aseptic filling line.

1. Prior to scheduled 6 months intervals in case of
   - Facility modification
   - Equipment modification
   - Change in equipment/personnel
   - Anomalies in end product sterility testing
   - Anomalies in environmental testing results

51.3.3 Repeat Performance Requalification

As aseptic process or filling line shall be subject to repeat of performance qualification studies when

SOP No. QAS-032.00 Effective date: mm/dd/yyyy
Approved by:

- An action level exceeded, unless an assignable cause for the exceeded action level is identified.
- Production lines have not been in operation for an extended period of time, for example, more than 6 months.
- There has been significant changes such as:
  a. Modification in equipment directly contacting bulk product or product container.
  b. Modification in equipment/facilities, which can affect air quality.
  c. Major changes in production personnel.

51.3.4 Product/Container Configuration

Configuration representative of other filled products which bracketed all others with respect to size, fill, container opening, line speed, manipulations, and so on or the product deemed to be the worst case with respect to the opportunity for contamination, for example, a container which has the largest opening and is conveyed at the slowest line speed.

51.3.4.1 Selection of Container

51.3.4.1.1 Vials Products

Currently the ABC Pharmaceutical Company uses four sizes of vials, that is, 10, 15, 20, and 30 mL vials. To ensure the sterility during the aseptic processing, it was decided to simulate the media fill runs on only 10 and 30 mL vial sizes since the 15 and 20 mL vial sizes will be bracketed in between 10 and 30 mL vial sizes. A measure of 10 mL vials with maximum speed, minimum volume and 30 mL vials with minimum speed, maximum volume are taken as worst cases. The details are as follows:

Size/Description	10 mL	15 mL	20 mL	30 mL
Fill volume (mL)	8	10	10	15
Closure diameter (mm)	20	20	20	20
Line speed	80%	80%	70%	50%
	160 vials/min	160 vials/min	140 vials/min	124 vials/min

In addition to 10 and 30 mL vials, media fill will also be carried out on 3 mL vial size using 90% or 180 vials/min line speed.

51.3.4.1.2 Ampoules

Currently the ABC Pharmaceutical Company is using three sizes of the ampoules, that is, 1, 2, and 3 mL ampoules for aseptically filled products. To ensure the sterility during the aseptic processing, it is decided to simulate media fill runs only for 1 mL ampoule and 3 mL ampoule sizes, since the 2 mL ampoule size will be bracketed in between 1 mL and 3 mL ampoule sizes. Ampoules of 1 mL will be run with maximum speed and minimum volume, whereas 3 mL ampoules with minimum speed and maximum volume will be taken as worst cases. The details are as follows:

Size/Description	1 mL	2 mL	3 mL
Fill volume (mL)	1	2	3
Line speed	100%	100%	92%
	200 Amp./min	200 Amp./min	184 Amp./min

SOP No. QAS-032.00 Effective date: mm/dd/yyyy
Approved by:

51.4 GENERAL REQUIREMENTS FOR ALL MEDIA FILLS

The following list details requirements for all aseptic process simulations:

a. A total of four people are to be present in the filling room (two operators, microbiologists, and technicians or maintenance) during the filling of the entire fill. Any additional personnel should be identified and documented (two operators during vial filling are necessary).
b. All activity carried out during filling (or lyophilization) should be as per the related SOPs, identified in the specified MFMs.
c. All personnel scheduled to work under aseptic conditions must participate in at least one media fill.
d. New employee must participate in the next scheduled media fill after receiving pertinent clean room training as per the appropriate SOPs.
e. Each semi-annual requalification run must be of a duration greater than/or equivalent to a routine fill and is to occur using both first and second filling shifts. The duration of biannual requalifications is a maximum of 24 hours.
f. The fill volume should not be less than approximately 50% of vials/ampoules.
g. Checking of the in-process fill is not allowed during media fill except during initial volume adjustment.
h. During each media fill, all major and minor defined interventions will be performed once. The time and the personnel performing the intervention will be documented as per the approved MFM.
i. The media fill operations shall be monitored for the operator's behavior by the operations manager and/or the manufacturing director (the usual and unusual intervention) through fixed cameras.
j. If multiple sizes of the same container/closure configuration are aseptically filled, selective sizes may be used for media fills. Containers with the widest diameter openings and slowest line speed should be included in the media-filling regimen and may be representative of worst-case conditions. However, the small containers are representative of worst conditions because of lack of container stability in the line operations.
k. Clear glass vials/ampoules should be used for media fill runs.
l. The duration of the run should be sufficient to cover all manipulations that are normally performed in actual processing, and all production shifts for line/product/container combinations.

51.4.1 Environmental Monitoring

Environmental monitoring is to be performed as per the current SOPs for the following:

A. Airborne bio-burden will be analyzed as per SOP, QCS-111.
B. Airborne particulate will be analyzed as per SOP, QAS-222.
C. Surface bio-burden will be analyzed as per SOP, QCS-333.
D. Personnel bio-burden will be analyzed as per SOP, QCS-444. In addition, QC/microbiology will perform audit monitoring of all media fill participants including themselves.

SOP No. QAS-032.00 Effective date: mm/dd/yyyy
Approved by:

51.5 DATA REVIEW

- Manufacturing will compile/review the batch documents and sent to QA.
- QC will examine and send the environmental monitoring and media fill vial/ampoule reports to QA.
- All data will be completed and reviewed within 21 days after the completion of the simulation of the process.
- QA will review the media fill batch records as per SOP, QAS-111 and issue the final report.

51.6 MEDIA FILL REPORT/DOCUMENTATION

A. For each media fill qualification a summary report will be written including the following information:

1. Reason for the media fill.
2. Type of media filled.
3. Date, time duration, and lot number of the media fill.
4. Filling room (re) qualification.
5. Names of the participants in the media fill.
6. Number and size of vials and/or ampoules filled as well as the type and size of closure used.
7. Target volume of media used (volume filled per container).
8. Machine speed of the vial/ampoules filling.
9. Container/closure type and size.
10. Filter lot and catalogue number.
11. Number of units rejected and reason (vials without stoppers, ampoules failed in leak test and initial volume adjustments only).
12. Reconciliation of the batch at each step with signature and date.
13. Number of units incubated.
14. Number of units positive.
15. Incubation time and temperature for each group of units incubated and whether any group of units is subjected to two different temperatures during the incubation.
16. Procedures used to simulate any steps of a normal production fill.
17. Microbiological monitoring data obtained during the media-fill set-up and run.
18. Growth promotion results of the media removed from filled containers.
19. Identification of the microorganisms from any positive units and investigation of any contamination events observed during media fills.
20. Length of time media was stored in holding tank prior to filtration.
21. Length of time taken to fill all containers.
22. Results of the requalifications as outlined in the acceptance criteria of the SOP.
23. Verification of media sterility.
24. Media fill unit inspection (records and reports).
25. Management review.

SOP No. QAS-032.00 Effective date: mm/dd/yyyy
Approved by:

51.7 PROCEDURE

51.7.1 Solutions

51.7.1.1 Media to Be Used

Soybean casein digest (TSB) broth shall be used for normal media runs while thioglycolate (TGB) broth should be used for detection of anaerobic organisms when using a filtered nitrogen purge to ensure anaerobic conditions.

Growth promotion batch of media should be challenged before using the media.

Prepare the media according to manufacturing direction, adjust pH, and sterilized by using 0.2 μm sterile filtration technique.

51.7.1.2 Manufacturing Methods/Process Simulation

The manufacturing methods/processes simulation include the following steps; refer to individual manufacturing formula & method (MFM) for each media fill run.

Filling equipment preparation	Freeze drying process (if required)
Media preparation	Vials transport and loading of freeze dryer
Sterilization	Vials incubation in FD chamber under vacuum
Sterile filtration	
Washing of commodities	Vials stoppering, unloading, and sealing
Depyrogenation	
Media addition	
Filling	

The filling operation of nonlyophilized products should be completed within 12 h (two shifts) as shown in Table 51.1.

51.7.1.3 Verification of Medium Sterility

Perform sterility test for the bulk media after filtration while bio-burden count before filtration.

51.7.1.4 Filling Operations

1. The containers and closures should be cleaned and sterilized as per current SOPs.
2. The filling machine is operated at the predetermined and validated fill rate for the used container size.

TABLE 51.1
Nonlyophilized Products

Runs	Filling 12 h		Intervals/Break
Run I	Shift 1/Day 1	Shift 2/Day 1	3–4
Run II	Shift 1/Day 2	Shift 2/Day 2	3–4
Run III	Shift 1/Day 3	Shift 2/Day 3	3–4

SOP No. QAS-032.00 Effective date: mm/dd/yyyy
Approved by:

3. Volume check is done only during machine setting.
4. During filling all major and minor interventions will be performed as defined in approved MFM and recorded.
5. Power shutdown to be simulated by stopping the laminar flow unit and the main AHU's for 30 s.
6. The containers are sealed and the media-filled units are collected in sequentially numbered trays or boxes (to correlate to the filling time).
7. The filled units should be briefly inverted and swirled after filling to assure closure contact with the medium.
8. Increase the size of the filling crew to more than the number necessary to fill the batch (maximum three people).
9. All routine activities which take place on the filling line should be a part of the test, that is, weight adjustments (only initial machine setting allowed), stoppers loading, and so on.

51.7.2 Lyophilized Products

51.7.2.1 Compounding and Filling Operations: As Above

Compounding as described in Section 51.7.1.

Filling operations will also be carried out as mentioned above but the containers are to be partially stoppered, which will be transferred through the dedicated conveyors to the automatic loading/unloading systems (ALUS).

The filling operation of lyophilized products shall be accomplished as indicated in Table 51.2.

51.7.2.2 Lyophilization Simulation

Media fill test should simulate the normal production environmental condition and process flow, including filling of product, loose capping, covering, transport to a freeze dryer, uncovering, filling of freeze dryer vacuum process, and seating of stoppers.

During the process, manual intervention may occur, including removal of tray tops or bottoms in the chamber, when loading vials, connections of thermocouples or thermoprobe junctions and unloading the chamber. All manual intervention should be included during scheduled media fill.

The freeze dryer media fill should be duplicated as much of the freeze drying process except freezing, as practical, including partial vacuum and duration as appropriate for the media. The methods employed for the lyophilization process simulation testing generally is similar to those used for solution fills with the addition of the transport and freeze-drying steps.

However, it should be focused on loading and sealing activities, which are presumed to be the greatest source of potential contamination. The test will be conducted according to the process details and description in the MFM.

TABLE 51.2
Lyophilized Products

Runs	Filling and Loading	Unloading
Run I	Day 1/Shift 1 with interval	On day 3
	Day 2/Shift 1 with interval	
Run II	Day 1/Shift 1 with interval	On day 3
	Day 2/Shift 1 with interval	
Run III	Day 1/Shift 1 with interval	On day 3
	Day 2/Shift 1 with interval	

SOP No. QAS-032.00 Effective date: mm/dd/yyyy
Approved by:

Containers are filled with media, and stoppers are partially inserted. The containers are loaded into the lyophilizer. A partial vacuum is drawn on the chamber and this level is held for a predetermined time. The vacuum must not be so low as to permit the medium in the containers to boil over. The chamber is then vented and the stoppers are seated within the chamber. The stopper units are removed from the aseptic processing area and sealed.

Anaerobic condition: If there is a need for sterile inert gas to break the vacuum on the chamber and remain in the container after sealing. The use of anaerobic media (e.g., alternative fluid thioglycolate medium) would be appropriated where the presence of anaerobic organisms has been confirmed in either environmental monitoring or, more likely, during end product sterility testing.

When anaerobic have not been detected in the environmental monitoring or sterility testing, lyophilizer process simulation tests should utilize TSB and air.

Lyophilizer media fills will consist of a simulation of all filling and handling procedures that would be required when processing a lyophilized product, such as, steam sterilization of lyophilizer 12 h (minimum) before loading the first tray of media, introduction of the vials into the lyophilizer, holding vacuum for a minimum of 24 h and subsequent releasing of vacuum and activation of the stoppering mechanism.

51.7.2.3 Media Filled Vials/Ampoule Analysis

Media filled units will be incubated in an upright position for 21 days at two temperature ranges. The units will be inspected for growth at two separate times, once at a minimum of 7 days of incubation at 22.5 ± 2.5°C and 14 days of incubation at 32.5 ± 2.5°C. Record of these inspections will be documented in QC registers. Any unit found which are positive for growth will be investigated as detailed in the Acceptance Criteria section of the SOP.

51.7.2.4 Growth Promotion of Media-Filled Units

Growth promotion of the media filled units will be performed on representative units of each media fill run after the completion of 21 days incubation period per instruction given in the Standard Test Method MC-000. The media should be challenged after subjecting to the same condition of the other media fill units.

Growth promotion ability of the medium in final filled containers must be demonstrated using 10 randomly selected containers for each challenge organism (more than 100 microorganisms per test container).

The units used for growth testing must be subjected to the same processing steps (e.g., cleaning, depyrogenation, sterilization, filtration, filling, lyophilization, and reconstitution) up to the point at which they are placed in an incubator.

In case of growth promotion controls, growth must be observed in at least 95% of the test containers for all the challenge microorganisms.

If no growth is observed in all of the 10 challenged containers, a second stage test must be conducted to rule out laboratory. In the second stage test both containers must support growth.

51.7.2.5 Negative Control

- Media from the bulk media container is incubated at a minimum of 7 days of incubation at 22.5 ± 2.5°C and at 32.5 ± 2.5°C for 14 days as negative control for liquid media fill to check its sterility.
- No growth should be demonstrated in negative controls to indicate the acceptability of media fill. If growth is observed the discrepancy should be investigated and documented.

SOP No. QAS-032.00 Effective date: mm/dd/yyyy
Approved by:

51.7.2.6 Incubation and Inspection of Media-Filled Units

- Leaking or damaged media-fill evaluation units shall be removed, and a record made of such removal, following processing and prior to incubation of the media.
- Media-filled evaluation units shall be incubated for a minimum of 21 days.
- Incubation temperatures shall be appropriate for the specific growth requirements of microorganisms that are anticipated in the aseptic filling area.
- Media-filled units should be stored or manipulated to allow contact of the media to all product contact surfaces in the unit.
- After completion of the incubation period, the media-filled containers shall be visually inspected for the presence of microbial growth.
- Microorganisms present in contaminated units shall be identified to help determine the likely source of the contamination.

51.7.2.7 Acceptance Criteria

Acceptance criteria for initial performance qualification and requalification of an aseptic processing line are shown in Table 51.3.

a. Successful USP growth promotion test results must be provided for the media utilized in the media fill.
b. Any contaminated unit(s) (not just failures) will be investigated. This investigation should include, but is not limited to, the following:
   1. Any recovered microorganisms shall be investigated related to the reason and possible origin of the recovered microorganism(s).
   2. Examine contaminated unit(s) for a breach of its integrity.
   3. Identify positive unit growth to at least the species level.
   4. Examine the activities occurring during the filling of the unit in the tray where the positive vial was found.
   5. Review sanitization records.
   6. Review environmental monitoring data including personnel monitoring data and training records. Alert and action level results will also be documented as per SOP, QCS-000.

TABLE 51.3
Alert and Action Levels

Number of Units in Single Media Fill Test	Alert Level (Number of Contaminated Units in a Single Media Fill Run)[a]	Action Level (Number of Contaminated Units in a Single Media Fill Run)[b]
≤5000	Nil	1*
5000–10,000	1**	2***
>10,000	1****	2***

[a] If a contaminated unit equals the alert limit: Investigate cause, conduct one additional run.
[b] If a contaminated unit is less than the action limit: Cease qualification media fill, investigate cause and repeat initial performance qualification (three media fills).
*One contaminated unit is considered the cause for revalidation, following an investigation.
**One contaminated unit should result in an investigation, including consideration of a repeat media fill.
***Two contaminated units are considered the cause for revalidation, following an investigation.
****One contaminated unit should result in an investigation.

SOP No. QAS-032.00 Effective date: mm/dd/yyyy
Approved by:

51.7.2.7.1 Media-Fill Runs Exceeding Alert Levels and Corrective Actions

Perform investigation as follows:

When media-fill action levels are exceeded, an investigation shall be conducted and documented regarding the cause.
If action levels are exceeded, there shall be a prompt review of all appropriate records relating to aseptic production between the current media fill and the last successful one.

The investigation should include, but not be limited to, consideration of the following:

a. Microbial environmental monitoring data.
b. Particulate monitoring data.
c. Personnel monitoring data (finger impressions, etc.).
d. Sterilization cycles for media, commodities, and equipment.
e. HEPA filter evaluation (airborne particulate levels, smoke-challenge testing, velocity measurements, etc.).
f. Room air flow patterns and pressures.
g. Operator technique and training.
h. Unusual events that occurred during the media fill.
i. Storage conditions of sterile commodities.
j. Identification of contaminants as a clue to the source of the contamination.
k. Housekeeping procedures and training.
l. Calibration of sterilization equipment.
m. Pre- and postfilter integrity test data, and/or filter housing assembly.
n. Product and/or process defects, and/or limitation of inspectional processes.
o. Documented disqualification of samples for obvious reasons prior to final reading.
p. Critical systems (HVAC, compressed air, gas, water, and steam) should be reviewed for documented changes.
q. Calibration records should be checked.
r. All HEPA filters in the filling area should be inspected and rectified, if warranted.
s. Training records for all individuals (production, maintenance, cleaning) involved in the fill should be reviewed to assure that proper training was provided.
t. If the root cause is assignable, corrective action needs to be taken and documented.

51.7.2.8 Corrective Actions

- Media-fill tests which exceed alert levels shall require action as described in Table 51.3.
- Decisions on whether or not to take action against the product being held and/or distributed shall be based upon an evaluation of all the information available and shall be documented.
- All product(s) that has been produced on a line following the media fill shall be quarantined until a successful resolution of the media fill has occurred.

Note: A review of production batches in association with an unsuccessful media fill should include appropriate environment monitoring data the record of sterility test results over this period, possible assignable causes for current media fill results, and any other information that would bear upon the sterility of the product involved.

SOP No. QAS-032.00 Effective date: mm/dd/yyyy
Approved by:

51.7.2.8.1 Media-Fill Runs Exceeding Action Levels and Corrective Actions

Investigation shall include all the steps defined in Section 51.7.2.7.1.

- A contaminated container should be examined carefully for any breach in the container system.
- All positives (from integral containers) should be identified till the species level.
- The identification of contaminant should be compared to the database of the organisms recently identified.
- The biochemical profile of the contaminant can be compared with that of microorganisms obtained from the sterility tests and bio-burden and environmental monitoring programs, in order to help identify the potential sources of the contaminant.

If the media fill contaminant is the same as the sterility test contaminant:

- Increase media fill units and routine filling environmental monitoring to identify the source of contamination.
- Review environmental data obtained during line set up.

If the media fill contaminant is the same as the media fill environmental contaminant:

- Increase routine environmental monitoring to determine if the contamination potential exists during routine filling operation.

If the media fill contaminant is the same as the routine environmental contaminant:

- Increase media fill environmental monitoring (in the same location) to confirm the contaminant source.

If the sterility tests contaminant is the same as the media fill environmental contaminant:

- Increase routine environmental monitoring (in the same location) and number of media fill vial to conform.

If the sterility test contaminant is the same as the routine environmental contaminant:

- Sterility test voided.
- Investigate sterility test procedures and room sanitation/sterilization methods to eliminate cause.

If the media fill environmental contaminant is the same as the routine environmental contaminant:

- Increases the number of media fill vials in media fill in order to determine the product risk potential.
- Review monitoring technique for possible problem.
- Review personnel practices, gowning, sanitation, and sterilization.

If the media fill environmental contaminant is the same as sterility test contaminant and if the routine environmental contaminant is the same as the sterility test contaminant:

- Check environmental monitoring methods and techniques closely for problems.
- Review personnel practices, gowning, sanitation, and sterilization.
- If the contaminated unit *is less* than the action limit: Cease qualification media fill, investigate the cause, repeat initial performance qualification (three media fills).

51.8 REASON FOR REVISION

First time issued for the ABC Pharmaceutical Company.

51.9 ATTACHMENT FORMS

ABC PHARMACEUTICAL COMPANY

QUALITY CONTROL LABORATORY

SOP: QAS-032
Issue Date:
LEVEL-3

Attachment-A

ENVIRONMENTAL MONITORING

AREA PARTICULATES COUNT

Date: ________________

Frequency: Daily **Building "C"**

ROOM NO.	ACTIVITY	CLEANNESS CLASS	LOCATION	PRODUCT/BATCH IN-PROCESS	RESULTS
	Syringe pass room (LFH)	100	Under LFH		
	Syringes filling (LFH)	100	Under LFH		
	Loading/unloading autoclave	100	Under LFH		
	Vials filling (LFH)	100	Under LFH		
	Ampoules filling (LFH)	100	Under LFH		
	Load/unload freeze dryer (LFH)	100	Under LFH		
	CR/1 unload FD (LFH)	100	Under LFH		
	Sterility testing (LFH)	100	Under LFH		

Tested by: ____________________ Date: ____________________
Checked by: ____________________ Date: ____________________

____
Signature
QA officer

Corrective action taken in case of failure

____
____
____
____
____
____
____

CC: Production manager
Maintenance engineer (Utilities)
QA manager (in case of failure)
Plant director (in case of failure)

ABC PHARMACEUTICAL COMPANY

QUALITY CONTROL LABORATORY

SOP: QAS-032
Issue Date:
LEVEL-3

Attachment-B

ENVIRONMENTAL MONITORING

AREA PARTICULATES COUNT

Date: ______________

Frequency: Twice/Week **Building "C"**

ROOM NO.	ACTIVITY	CLEANNESS CLASS	LOCATION	PRODUCT/BATCH IN-PROCESS	RESULTS
C-06	Syringe preparation	10000	Central of the room		
C-07	Change room-I (at rest)	10000	Center of the room		
C-08	Change room (at rest)	100	Center of the room		
	Change room-II (at work)	10000	Center of the room		
C-12	Syringes filling remaining room	10000	Center of the room		
C-18	Air lock for materials	10000	Center of the room		
C-19/1	Personnel air lock–II	10000	Center of the room		
C-19/2	Solution preparation	10000	Center of the room		
C-20	Vials capping (LFH)	10000	Center of the room		
C-25	Vials/ampoules filling remaining room	10000	Center of the room		
C-28	CR/1 Unload FD	10000	Center of the room		
C-31	Solution room (press box)	10000	Center of the room		

Tested by: ______________ Date: ______________
Checked by: ______________ Date: ______________

____
Signature
System & inspection officer

Corrective action taken in case of failure

____
____
____
____
____
____
____

CC: Production manager
Maintenance engineer (Utilities)
QA manager (in case of failure)
Antibiotic plant director (in case of failure)

ABC PHARMACEUTICAL COMPANY

QUALITY CONTROL LABORATORY

SOP: QAS-032
Issue Date:
LEVEL-3

Attachment-C

ENVIRONMENTAL MONITORING

AREA PARTICULATES COUNT

Date: ________________

Frequency: Once/Week **Building "C"**

ROOM NO.	ACTIVITY	CLEANNESS CLASS	LOCATION	PRODUCT/BATCH IN-PROCESS	RESULTS
C-04	Production corridor	100000	Opposite autoclave opposite FD Tech.R opposite Pack. A		
C-05	Syringe remaining room	100000	Center of the room		
C-07	Change room-I (at work)	100000	Center of the room		
C-09	Coding room	100000	Center of the room		
C-10	Air lock washing	100000	Center of the room		
C-11	Packaging A	100000	Center of the room		
C-17	Washing area for compounding vessels	100000	Center of the room		
C-19	Personnel air lock-I	100000	Center of the room		
C-22	Commodity washing room Autoclave loading	100000 100000	Near loading Near loading		
C-23	Packaging B	100000	Center of the room		
	Sterility testing room	100000	Center of the room		

Tested by: ____________________ Date: ____________________
Checked by: ____________________ Date: ____________________

____
Signature
System & inspection officer

Corrective action taken in case of failure

____
____
____
____
____
____
____

CC: Production manager
Maintenance engineer (Utilities)
QA manager (in case of failure)
Antibiotic plant director (in case of failure)

ABC PHARMACEUTICAL COMPANY

SOP No.: QAS-032
ISSUED DATE:

QUALITY CONTROL LABORATORY
ENVIRONMENTAL MONITORING
AIR SAMPLING
ANNEXURE I

ROOM	SAMPLING STATION	CLEANLINESS	PRODUCT/ BATCH	LIMIT CFU/CF OF AIR	RESULT
	Cartoning	100000			
	Bottling	10000			
	Capsule	10000			
	Tabletting	10000			
	Store	100000			
	Washing	100000			
	Blending saving	10000			
	Weighing	10000			
	QA	100000			
	Change room	100000			
	Washing	100000			
	Coating	10000			
	Sorting	10000			
	Blistering	100000			
	Pre-airlock	100000			
	Change room/1	100000			
	Change room/2	10000			
	Solution preparation	10000			
	Change room (2)	100000			
	Loading bottle	10000			
	Loading capsule	10000			
	Loading tablet	10000			
	Production corridor	100000			
	Loading tablet	100000			
	QC laboratory	100000			

Prepared by: ____________________ Date: ____________________
Tested by: ____________________ Date: ____________________
Checked by: ____________________ Date: ____________________

____
QC manager

Corrective action taken in case of failure:

____
____
____

CC: QA manager/production manager/maintenance engineers (Utilities)

ABC PHARMACEUTICAL COMPANY

SOP No.: QAS-032
ISSUED DATE:

QUALITY CONTROL LABORATORY
ENVIRONMENTAL MONITORING
SEDIMENTATION PLATES
ANNEXURE II

ROOM	SAMPLING STATION	CLEANLINESS	PRODUCT/ BATCH	LIMIT CFU/CF OF AIR	RESULT

Prepared by: ____________________ Date: ____________________
Tested by: ____________________ Date: ____________________
Checked by: ____________________ Date: ____________________

____
QC manager

Corrective action taken in case of failure:

____
____
____

cc: QA manager/production manager/maintenance engineers (Utilities)

LEVEL 3
SOP No.: QAS-032
Issued on:
Revision No.:
Initiator name:

QUALITY CONTROL DEPARTMENT

Environmental Monitoring

Media Fill Test

Test Date: / /

Reason for test: ....

Product simulated for: ....

Production procedure: ....

Media fill procedures: ....

No. and names of operating personnel: ....

....

....

....

Production facilities:

Item	Identification	Status
Preparation tank		
Mobile tank		
Filtration unit		
Holding tank		
Filling machine		
Small utilities		
Autoclave		
Tunnel		

Batch (media) size: ....

Number of units to be filled: ....

Number and type of interventions and stoppages to be included in the test

....

....

The rooms to be used: ....

Volume of medium to be filled in the containers: ....

....

Preparation time: ....

....

LEVEL 3
SOP No.: QAS-032
Issued on:
Revision No.:
Initiator name:

Hold time (if any): ....

....

Type of container/closure to be used: ....

....

Line speeds

Low ....
Normal ....
High ....

Media or placebo materials to be used: ....

Environmental reports of the system during the test: ....

No. box and tray: ....

Remarks on the media fill process: ....

....

....

Verification of medium sterility: ....

....

Result of growth promotion test: ....

....

Incubator identification: ....

Incubation time: ....

Incubation temperature: ....

....

Results	Time Inspection of Filled Units			
	1st Day	7 Days	10 Days	14 Days
Number of units incubated				
Number of units rejected (damaged container)				
Number of units rejected (defective seal container)				
Number of units positive				

Interpretation of results:

$$\textbf{The \% of contamination} = \frac{\textbf{No. of vials with microbial growth} \times 100}{\textbf{No. of vials filled} - \textbf{No. of damaged vials}}$$

= %

The acceptable percentage of contamination is ≤ 0.1%.

LEVEL 3
SOP No.: QAS-032
Issued on:
Revision No.:
Initiator name:

Reports about media fill environmental contaminant:

....

....

The identification of contaminant

Method: ....

....

Name of the contaminant(s): ....

....

The biochemical profile of the contaminant: ....

....

....

....

Comparison with microorganisms obtained from the sterility tests: ....

....

....

Comparison with microorganisms obtained from media fill environmental contaminant:

....

....

Comparison with microorganisms obtained from routine environmental contaminant:

....

....

Summary of the comparison:

....

....

Corrective Action

In Case of Failure			Respond		Retest	
Media Fill No.	Contamination Comparison	Corrective Actions	Yes	No	Result	Remark
	Media fill contaminant same as sterility test contaminant	Increase media fill vial quantities and routine filling environmental monitoring to identify the source of contamination. Review environmental data obtained during line set up.				
	Media fill contaminant same as media fills environmental contaminant	Increase routine environmental monitoring to determine if the contamination potential exists during routine filling operation.				

(continued)

In Case of Failure			Respond		Retest	
Media Fill No.	Contamination Comparison	Corrective Actions	Yes	No	Result	Remark
	Media fill contaminant same as routine environmental contaminant	Increase media fill environmental monitoring (in the same location) to confirm the contaminant source.				
	Sterility test contaminant same as media fill environmental contaminant	Increase routine environmental monitoring (in the same location) and number of media fill vial to conform.				
	Sterility test contaminant same as routine environmental contaminant	Sterility test voided. Investigate sterility test procedures and room sanitation/sterilization methods to eliminate cause.				
	If media fill environmental contaminant same as sterility test contaminant and if routine environmental contaminant same as sterility test contaminant	Check environmental monitoring methods and techniques closely for problems. Review personnel practices, gowning, sanitation, and sterilization.				
	Media fill environmental contaminant same as routine environmental contaminant	Increases the number of media fills vial in media fill in order to determine the product risk potential. Review monitoring technique for possible problem. Review personnel practices, gowning, sanitation, and sterilization.				
Hold the Production						
		Critical systems (HAVAC, compressed air/gas, water, steam) should be reviewed for documented changes.				
		All HEPA filters in the filling area should be inspected and rectified, if warranted.				
		Training records for all individuals (production, maintenance, cleaning) involved in the fill should be reviewed to assure proper training was provided.				

Final comment:

....

....

....

Tested by: .... **Checked by:** ....
Date: .... Date: ....

Distribution:

1. Validation officer, production manager, maintenance engineer
2. In case of failure: QA manager, plant director

ABC PHARMACEUTICAL COMPANY

SOP No.: QAS-032
Issued on:
Revision No.:
Initiator name:

Attachment No. I
Media Fill Request Form

From : Manufacturing director/operations manager
To : List
Date :
Subject : **Initial/Periodic Performance Qualification**

Vials/Ampoules
Size: ________________

The media fill is scheduled from __________ to __________ to comply with the SOP, QAS-032 for the following reason.

Tick Applicable Box

1. Initial Performance Qualification

Three media fill runs are required for the following (this list is not all inclusive)

1.1 New filling room or machine ☐
1.2 Major room modification (such as wall or ceiling reconfigurations) ☐
1.3 New equipment or new machine which has product contact or affects product flow which is not an exact equivalent of the original equipment ☐
1.4 Major mechanical or line configuration changes (such as new equipment additions or major preventative maintenance) ☐
1.5 Major HVAC changes which may impact air flow patterns related to the critical filling area (Class 100) (such as addition/removal of HEPA housings) ☐
1.6 A media fill failure with a conclusive nonpersonnel assignable cause (determined during the investigation process) ☐
1.7 New product/process evaluation (number of runs to be determined on a case by basis) ☐

2. Periodic Performance Qualification

One media fill run is required for the following (this list is not all inclusive) ☐

2.1 Routine requalification (every 6 months) ☐
2.2 Major modification to areas adjoining filling rooms (such as expanded core area) ☐
2.3 Equipment modification ☐
2.4 Change equipment/personnel ☐
2.5 Anomalies in end product sterility testing/environment ☐
2.6 Any other reason: ____ ☐

3. Repeat Performance Qualification

As aseptic process or filling line shall be subject to repeat of performance qualification studies when:

3.1 An action level exceeded, unless an assignable cause for the exceeded action level is identified ☐
3.2 Production lines have not been in operation for an extended period of time, e.g., more than 6 months ☐

SOP No.: QAS-032
Issued on:
Revision No.:
Initiator name:

3.3 There has been significant changes such as: ☐
  a. Modification in equipment directly contacting bulk product or product container ☐
  b. Modification in equipment/facilities, which can affect air quality ☐
  c. Major changes in production personals ☐

4. No. of units to be filled ________________ MFM No. __________________

List: QA director/QC director/managers QA/microbiological section head

QAS-033

STANDARD OPERATING PROCEDURE

<table>
<tr><td rowspan="2">SUBJECT: Validity and Control of Standard Operating Procedures</td><td colspan="2">SOP NO.: QAS-033

REVISION NO.:</td></tr>
<tr></tr>
<tr><td>WRITTEN BY: DATE</td><td colspan="2">BASED ON:
Current Practice</td></tr>
<tr><td>CHECKED BY: DATE</td><td>SUPERSEDES</td><td>ISSUED ON:</td></tr>
<tr><td rowspan="3">APPROVED BY: DATE

AUTHORIZED BY: DATE</td><td>DISTRIBUTED TO:</td><td>Page 0 of 00</td></tr>
<tr><td colspan="2">EFFECTIVE DATE:</td></tr>
<tr><td colspan="2">REVIEWED DATE:</td></tr>
</table>

Subject: Validity and Control of Standard Operating Procedures

52.1 PURPOSE

To describe the procedure for writing, reviewing, approval, authorization, training, indexing, and distribution of standard operating procedures (SOPs) in production, packaging, QA, QC, stores, materials planning, research and development laboratory, maintenance, regulatory affairs, and purchase department.

52.2 SCOPE

The SOP QAS-033 is applicable to all standard quality procedures and guidelines.

52.3 RESPONSIBILITY

It is the responsibility of the departmental managers, supervisors, technicians, and operators to write SOPs related to their respective systems and equipments in coordination with area managers. The concerned departmental managers and the QA manager are responsible for the SOP compliance.

SOP No. QAS-033.00 Effective date: mm/dd/yyyy
Approved by:

52.4 PROCEDURE

52.4.1 SOPs Format

All SOPs will have the fixed format on each page (as above) apart from page 1 where the header of the SOP will be having additional space beside the "effective date" and "review date" to stamp the date when the particular SOP is becoming effective after the completion of SOP training evaluation and SOP validity date due for the next review, respectively.

52.4.2 Subject

In the space for "subject," write the heading of the SOP, for example, "Housekeeping of the plant."

52.4.3 SOP No.

All SOPs will be abbreviated according to the ABC Pharmaceutical Company numbers and department using the following alphabetical codes. Each abbreviation will be assigned a numerical number in an increasing order.

Example

SOP Code	Department	Number
QAS	Quality assurance system	001
ENS	Engineering system	001

Therefore, the number assigned to the SOP will be QAS-001 and/or ENS-001.

52.4.4 Written by

Preferably written by the systems follower and machine operator as required.

52.4.5 Checked by

All SOPs shall be checked by the departmental managers.

52.4.6 Approved by

All SOPs shall be approved by the QA managers or authorized by the technical manager.

52.4.7 Revision No.

In the space for "Revision No.," write the number, for example, 1 if the SOP is revised for the first time. If the SOP is written for the first time, do not assign any number. Instead write 00 in the space.

52.4.8 Based on

In the space for "Based on," write down the reference (e.g., CFR, WHO, or PICs) or preferably the current practice. It is the responsibility of the QA manager to identify and prepare a list of SOPs

SOP No. QAS 033.00 Effective date: mm/dd/yyyy
Approved by:

required based on cGMP requirements as PICs, WHO, EU, and US FDA guidelines in coordination with the concerned departmental managers.

52.4.9 Supersedes

In the space for "Supersedes," write "New" if the SOP is written for the first time or write the previous "Issued on" date of the previous SOP, if the SOP is revised.

52.4.10 Issued on

In the space for "Issued on," the QA manager will provide the date with 1 week margin from the date of writing. Within this 1 week period, the SOP will be checked and approved.

52.4.11 Review Date

In the space for "Review Date," write the SOP review date. SOPs review shall not exceed 30 days from the date of review. The (validity) review date for systems SOP is 24 months (2 years). The review date for fixed SOP is 60 months (5 years).

52.4.12 Distributed to

In the space for "Distributed to," enter the names of the departments where the SOP is applicable and will be used. To indicate the copy marked to, use preferably the position's abbreviation described below or, if applicable, use department/section's abbreviation under these two headings separately. The SOPs on display will also be marked in the "Distributed to" space.

The following table describes the positions and abbreviations to be used for SOP distribution and the list of samples, signatures, and initials of the employees responsible for writing, reviewing, approving, and authorizing the standard operating procedures.

Position	Position Abbreviation	Name	Signature	Initials
1. General manager	GMR	________	________	________
2. Technical manager	TM	________	________	________
3. Quality assurance manager	QAM	________	________	________
4. Marketing manager	MKM	________	________	________
5. Production manager	PRM	________	________	________
6. Maintenance manager	MM	________	________	________
7. Regulatory affairs manager	RAM	________	________	________
8. Quality control manager	QCM	________	________	________
9. Stores manager	STM	________	________	________
10. Planning manager	PMG	________	________	________
11. Packaging manager	PKM	________	________	________
12. Quality assurance in-process	QAI	________	________	________
13. Packaging officer	PKO	________	________	________
14. Purchase manager	PRM	________	________	________
15. Manager safety and security	MSS	________	________	________

SOP No. QAS-033.00 Effective date: mm/dd/yyyy
Approved by:

52.4.13 Page

In the space for "page," write the page number, for example, if the total pages is 5, write 1 of 5, 2 of 5, 3 of 5, 4 of 5, 5 of 5 in the ascending order.

52.4.14 Effective Date

The date will be provided by the QA manager after the validation and training of SOP by the concerned SOP writer. The effective date will be stamped only on the first page in the space provided.

52.4.15 SOP Writing/Checking and Approval

Each SOP will be written by the concerned employee and will be checked by the departmental manager before final approval by the QA manager.

52.4.16 SOP Write-up, Review, and Approval

It is the responsibility of all the concerned departmental managers to organize the writing of the related standard operating procedure and follow a fixed format as per the ABC Pharmaceutical Company's policy. All SOPs shall be counterchecked by the second person to verify the accuracy of the contents. All SOPs shall be approved by the QA manager.

52.4.17 SOP Training Evaluation/SOP Implementation (Effective Date)

After the QA approval, the master SOP will be copied by the QA manager and stamped as "uncontrolled copy" (in red color) to be forwarded to the SOP trainer accompanied with the standard operating procedure validation/training record form (Attachment 1) and SOP evaluation questionnaire for the systems only. The fixed SOP shall be subjected to training through SOP reading and on-site demonstration (e.g., Attachment 2).

52.4.18 SOP Training Evaluation Questionnaire

It is the responsibility of the SOP writer to prepare the SOP training questionnaire for each SOP and forward it to the QA manager (e.g., Attachment 2) at the time of first initiation and on consecutive revisions.

Uncontrolled Copy
For Training Only
Signature: ___________ Date: ________

The trainer will return the uncontrolled SOP with the executed training record form along with the evaluation questionnaire within 5 working days back to the QA manager.

52.4.19 SOP Distribution

1. The QA manager will attach the documented training record form and SOP evaluation questionnaire with the master SOP and stamp the effective date on the space provided in the front page of each SOP and the next review date, respectively.

SOP No. QAS 033.00 Effective date: mm/dd/yyyy
Approved by:

2. The "uncontrolled copy" will be destroyed in the shredder.
3. The QA manager will update the SOP index maintained on a computer with the following details:

Master SOP Index

Subject	SOP No.	Rev. No.	Supersedes Date	Issued Date	Effective Date	Review Date

4. The copies of the master SOP will be made as per the distribution list.
5. The SOPs distribution and retrieval record will be maintained in the SOP distribution record register (Attachment 3) with full traceability.
6. The master superseded SOPs will be stamped "void" (red color) and kept for an indefinite period in the QA archive.
7. The return superseded SOPs will be destroyed in the shredder.

52.4.20 Annual SOP Reading/Record

All new SOPs, revised SOPs, and current SOPs will be read once in a year by the concerned staff or explained to the concerned staff by the section in-charge regardless of the SOP training/evaluation questionnaire. The reading or explanation will be documented on the reading record register (Attachment 4) by the concerned manager. Not more than five SOPs should be read in a day.

52.4.21 Annual SOP Review

1. All systems SOPs shall be reviewed once every 12 months. And fixed SOPs will be reviewed every 60 months as per the corresponding site procedure.
2. The QA manager will plan the SOP review as per the information provided in the master SOP index.
3. The SOP will be forwarded to the person responsible for the review, except the QA systems to be reviewed by QC or production to avoid conflict of interest.
4. If the SOP does not require revision, the review form will be closed out and returned to the QA manager.
5. If the SOP requires revision, a separate change control form will be initiated as per SOP______, and closed-out after redistribution as per the existing standard operating procedure.

52.4.22 System SOP Validity

All system-related SOPs are valid for a period of 2 years + 30 days for revision.

52.4.23 Fixed SOP Validity

1. All procedures that are related to the fixed equipments and are highly unlikely to be changed frequently will have 5 years SOP review date + 30 days for revision.

SOP No. QAS-033.00 Effective date: mm/dd/yyyy
Approved by:

For example:

- Equipment operating procedures
- Calibration procedures
- Maintenance procedures
- Cleaning procedures
- Housekeeping procedures

52.4.24 Ongoing SOP Revision

1. The system-related SOPs and fixed SOPs can also be subject to review based on the comments from the end users, cGMP requirements, changes in hardware and software, comments from the internal and external auditors that shall follow change control SOP No.____________.

52.4.25 Display SOPs

The SOPs on display or location will be marked in the "distributed to" space as given below:

1. Production department/stores/weighing area, for example:
   a. Room No. 1 = R1 (DISPLAY)
   b. Room No. 1 and 2 = R1 and 2 (DISPLAY)
   c. Room No. 1 to 10 = R1–10 (DISPLAY)
2. Packaging department, for example:
   a. Line No. 1 = L1 (DISPLAY)
   b. Line No. 1 and 2 = L1 and 2 (DISPLAY)
   c. Line No. 1 to 14 = L1–14 (DISPLAY)
3. Quality control, for example:
   a. Microbiological laboratory = Micro. Lab.
   b. Physical laboratory = Phy. Lab.
   c. Chemical laboratory = Chem. Lab.
   d. Instrumental laboratory = Inst. Lab.
4. Product development lab, for example:
   a. Formulation
   b. Stability
   c. HPLC
   d. Mfg.
5. Maintenance, for example:
   a. Equipment operating procedures (as appropriate) on display in departments will be marked as (DISPLAY).
6. The SOP displayed at all places or lines will be marked as DISPLAY. All SOPs on display will also be logged in the SOP distribution record register (Attachment 3).

SOP No. QAS 033.00 Effective date: mm/dd/yyyy
Approved by:

7. The SOPs for the display will be photocopied on colored sheets as described below:

a. Equipment operating procedures	Green sheets
b. Equipment calibration procedures	Blue sheets
c. Equipment maintenance procedures	Blue sheets
d. Equipment cleaning and sanitation procedures	Pink sheets
e. Room cleaning and sanitation procedures	Yellow sheets
f. System-related procedures	White sheets
g. The display SOPs will be laminated. If lamination sheets are not available, they will be kept in the index folder and displayed.	

52.4.26 Confidentiality

Photocopies of all SOPs will be stamped in red color with "uncontrolled copy" except reporting formats.

52.4.27 Annexures and Reporting Formats

Annexures and reporting formats made part of the SOP will bear SOP No., issued of date, revision number, and initiator signature or name for traceability. The formats used to capture the raw data will be issued by the QA manager. All annexures and reporting formats made part of the SOP will be marked to describe their level in the documentation system.

52.4.28 Flow of SOPs

The QA manager will coordinate to keep the flow of drafted SOP to the appropriate designated writer for reevaluation and comments. After a SOP has been approved, it will come to the QA manager for distribution. The QA manager or designate is the only authorized person permitted to remove and make copies of procedures from the master files for distribution.

52.4.29 SOPs Indexing

The QA manager or the designated supervisor will maintain the approved SOPs index on an ongoing basis. Every quarter and prior to the regulatory and ISO inspection, the updated SOP index should be physically printed, signed off, and distributed by the QA manager to the departmental managers for verification.

Before all internal and external audits, the SOPs index shall be printed and provided to the auditor if requested.

52.5 REASON FOR REVISION

First time issued for the ABC Pharmaceutical Company.

52.6 ATTACHMENT FORMS

ABC PHARMACEUTICAL COMPANY

SOP No.:
Issued on:
Revision No.:
Initiator name:

Attachment 1
STANDARD OPERATING PROCEDURE VALIDATION/TRAINING RECORD

SOP No.: ______ Revision: ______ Issued on: ______

Trainees (Name)	ID No.	Signature of Understanding	Date	Evaluation	Results	Training Conducted by	Signature of the Trainer

Evaluation: O = Oral
R = Written
Q = Questionnaire

Results: M = Marginal (retraining required)
A = Average
S = Satisfactory
C = Commendable

ABC PHARMACEUTICAL COMPANY

SOP No.:
Issued on:
Revision No.:
Initiator name:

Attachment 2
SOP TRAINING QUESTIONNAIRE

Trainee: ______________________ **ID:** ______________

Trainer: ______________________ **ID:** ______________

Subject: ______________________ **SOP No.:** __________

Multiple Choice: Mark ✓ infront of the right answer.

1. SOP QAS______ is applicable to
   a. Standard operating procedures ☐
   b. Standard test methods ☐
   c. Registration documents ☐
   d. QC certificate of analysis ☐

2. All master superseded SOPs shall be
   a. Destroyed ☐
   b. Copied and distributed to all departments ☐
   c. Kept by the production manager ☐
   d. Stamped "void" in red and archived in QA ☐

3. All systems SOPs shall be reviewed
   a. Once every 12 months (1 year) or as and when required ☐
   b. Every 2 years ☐
   c. Every 5 years ☐
   d. Never ☐

4. All equipment and fixed operation-related SOP shall be
   a. Reviewed every 3 months ☐
   b. Reviewed every 3 years ☐
   c. Reviewed every 60 months (5 years) ☐
   d. Reviewed every 5 years or as and when required ☐

5. All SOPs shall be reviewed within
   a. 2 months after the review date ☐
   b. 3 months after the review date ☐
   c. 30 days from the review date ☐
   d. All of the above ☐

6. SOPs for distribution are
   a. Copied from the master SOP ☐
   b. Copied, stamped "uncontrolled," documented in the SOP distribution record register ☐
   c. None of the above ☐

7. All employees can make a copy from the departmental SOP
   a. Yes ☐
   b. No ☐

8. All employees can make changes in their SOP by pen or pencil without QA approval and use them
   a. Yes ☐
   b. No ☐

Total number of questionnairs: 8
Total correct answers:

Evaluation matrix:

M = marginal: 80–89%
A = average: 90–94%
S = satisfactory: 95–99%
C = commendable: 100%

Note: Results shall be discussed by the trainer, results less than 94% require retraining.

Signature: ________________________________ **Date:** ______________________

ABC PHARMACEUTICAL COMPANY

SOP No.:
Issued on:
Revision No.:
Initiator name:

Attachment 3
Standard Operating Procedure
Distribution Record

From: ______________________ **To:** ___________________

Department: ____

Log Book ID: _______________________________________

ABC PHARMACEUTICAL COMPANY

SOP DISTRIBUTION RECORD

Index

SOP Description	SOP No.	Page No.

SOP DISTRIBUTION RECORD

Page No.

SOP No.: ____________________

Subject: ____________________

Revision No.	Copies to	Issued on	Superseded	Effective Date	SOP Review Date	SOP Received			Superseded SOP Returned		
						From	By	Date	From	By	Date

Revision No.	Copies to	Issued on	Superseded	Effective Date	SOP Review Date	SOP Received			Superseded SOP Returned		
						From	By	Date	From	By	Date

SOP No.:
Issued on:
Revision No.:
Initiator name:

Attachment 4
STANDARD OPERATING PROCEDURE READING RECORD

QUALITY ASSURANCE DEPARTMENT

From: ______________________ **To:** ___________________

Department: _______________________________________

Room No.: ____

SOP READING RECORD
Index

S. No.	Name	Job Title	Area	Page No.

SOP READING RECORD

Name: ____

Job title: ____

SOP No.	Revision No.	Reading		Remarks	Time Spent (Minutes)
		Date	Signature		

Index

L

M

O

P

U

V

W